COGENERATION OF STEAM AND ELECTRIC POWER

COGENERATION OF STEAM AND ELECTRIC POWER

Edited by Robert Noyes

NOYES DATA CORPORATION
Park Ridge, New Jersey, U.S.A.
1978

Library of Congress Catalog Card Number: 78-55528
ISBN: 0-8155-0706-2
Printed in the United States

Published in the United States of America by
Noyes Data Corporation
Noyes Building, Park Ridge, New Jersey 07656

Foreword

The cogeneration concept, as presented in this book, refers to the generation of both process steam and electricity on site by heating water with a primary fuel, such as coal or oil or gas.

When fuel was cheap, industries were not interested in the capital outlays necessary for cogeneration equipment. Now, however, the picture has changed. Refineries, chemical plants, paper mills and other big energy users can get maximum value from their fuel by a "topping cycle," viz. producing steam and running it through a turbine generator before using the steam for processing operations. Also, cogeneration can be accomplished through a "bottoming cycle," utilizing process steam prior to electrical power generation. Both methods are discussed in detail.

In the past few years, the U.S. Government has been placing much emphasis on the need for conserving energy. One of the ways of saving considerable amounts of energy is the development of cogeneration plants by those industries which use significant amounts of both electricity and thermal energy for process heat and/or space heating. Since electric power generation produces heat as a by-product, any thermal energy which can be used produces a fuel savings of from 10 to 30%, depending on the applications and methods involved. The studies indicate that for the greatest potential development of cogeneration, government support may be required.

Because the information in this book is taken from several different sources, it is possible that certain portions of this book may disagree or conflict with other parts of the book. This is especially true of monetary values and opinions of future potential. We chose to include these different points of view, however, in order to make the book more valuable to the reader. Cost figures provided are those given in the report cited, the date of which is always given. When the dates of the cost figures themselves are given, we have included them.

Advanced composition and production methods developed by Noyes Data are employed to bring these durably bound books to you in a minimum of time. Special techniques are used to close the gap between "manuscript" and "completed book." Industrial technology is progressing so rapidly that time-honored, conventional typesetting, binding and shipping methods are no longer suitable. We have bypassed the delays in the conventional book publishing cycle and provide the user with an effective and convenient means of reviewing up-to-date information in depth.

Various technological studies, such as material prepared for the National Science Foundation and the Federal Energy Administration, were the basis for this review. The bibliography at the end of the volume lists the reports of these studies and the table of contents (at the beginning) is organized to serve as a subject index and provides easy access to the information presented in this book.

Contents and Subject Index

Dow Chemical Company Study

Material for this chapter is taken from report PB-243 823, published in 1975, and prepared for the National Science Foundation by the Dow Chemical Company.

INTRODUCTION

Background

The present and future energy shortage has been well publicized of late. The time lags and other deterrents to rapid expansion of our domestic fuel production, coupled with the difficulties associated with substantial increases in oil imports, have made it abundantly clear that every significant route to more efficient use of our fuel resources must be considered in any program of national self-sufficiency in energy. Attention is being devoted to energy conservation in every sector of the United States socioeconomic structure.

Two major consumers of energy are industry and the electric utilities, which together account for well over half of all the primary fuel used in the country. The effectiveness of this utilization makes an impact on the final consumer in terms of the availability and cost of goods and electric power, and should thus be a general concern. Both industry and the utilities are being strongly affected by the scarcity of fuels presently considered acceptable and by rising prices.

The electric utilities, especially, are under heavy pressures from all sides. It has been estimated that $305 billion will be required during the next ten years to construct new plants for electrical generation. This is roughly equivalent to building new generating facilities each year of the decade capable of supplying the electrical needs of the cities of New York, Chicago, Los Angeles, Philadelphia, Detroit and Houston. At the same time, costs for generating electricity have been rising at a very rapid rate, because of the increasing cost of fuels and of construction, and in some parts of the country a virtual rebellion by consumers has resulted. The high visibility of the electric utilities makes them a convenient target for the venting of public frustration over widespread inflation.

Supplying this enormous demand and at the same time maintaining reasonable energy costs presents extraordinary demands on the utilities. Not the least of these is an increasing difficulty in financing needed expansion in the face of inadequate cash flows, high interest rates, declining bond ratings, and depressed stock prices. Clearly, any direction that offers the opportunity of reducing capital investment and fuel consumption while satisfying demand must be investigated.

In excess of 17% of the nation's primary fuels go to produce industrial steam. Wherever industrial steam is generated, there is potential opportunity for generating electricity also, although this is taken advantage of in only a minority of cases. It has been known for many years that raising the steam pressure and temperature above that required for industrial use and bringing it down to the desired conditions through a turbine coupled to a generator would produce electric power with about half the fuel consumption of the most efficient central-station power plants. Not all industrial boiler installations supply a large enough steam load to make electric generation economically attractive, but the unexploited potential is still very large.

Similar fuel savings are possible by supplying steam to industry from utility central-station power plants. Some such arrangements have been in effect for many years, and new ones are still being formed, but the sum of all of them represents a small fraction of the total possible.

Expanded adoption of these two approaches to supply industrial steam requirements could well have a significant impact on the national problems of fuel conservation and financing electric generation additions. It also raises questions of environmental impacts and legal and regulatory implications.

TECHNICAL AND INDUSTRIAL CONSIDERATIONS

Power and Steam in Industry

Historical Development: At the turn of the century, the majority of industrial plants generated their own utilities; in 1960, industry generated only 21% of the electricity it consumed, dropping to 17% by 1968 (1). A very high percentage of industrial organizations still generate their own steam, although during an intermediate period, the electric utilities sold substantial quantities. This separation of steam and power generation has eliminated a large part of the advantages to be found in combined operations.

A number of factors contributed to this split in utility production over a rather extended period. Technical progress encouraged the introduction of larger generating units with all the associated economies of scale. Fuel cost increases at the time of the two World Wars accelerated the move toward more efficient units. The resulting lower power costs stimulated the demand for electricity and vigorous competition for the expanding markets ensued, with a variety of accompanying abuses. This problem was resolved by the granting of franchises together with government regulation of the resulting monopolies, initiating a period during which capital flowed freely and the utilities enjoyed a high state of efficiency and rapid growth.

In the early years, many urban utilities supplied steam for district heating. The condensate was not ordinarily recovered because of the cost factors; indeed, the usual meter for a customer's steam bill simply measured the liquid condensate before dropping it into the sewer. This did not pose a major problem as long as boiler pressures were low enough to permit fairly simple water treatment for boiler feed. But the drive for more efficiency brought higher and higher boiler pressures and temperatures, and with them, the need for much higher water purity. If pure condensate could not be virtually completely recovered, the only options seemed to be to discontinue steam sales or install very elaborate and costly water treatment plants.

These and many other factors motivated the utility companies to concentrate on the more profitable sale of electricity. At this, they were most successful, and electric power rates decreased continually for many years, in spite of the rise in all other costs.

Meanwhile, a variety of factors were causing industrial managers to have second thoughts about the desirability of generating power for their own operations, including the following:

1) Corporate income tax rates that encouraged the substitution of expenses for capital investments, and generally tended to foster shorter-term thinking.
2) A difference in risk that necessitated a higher return on an industrial investment than that required by utilities.
3) Decreasing electric rates from central power stations that encouraged substituting power for steam in some situations, thereby eroding the industrial steam base.
4) Changes in the labor situation that made it difficult to find and keep people able to handle the problems of coal-fired boilers. Gas and oil-fired packaged boilers which could operate almost unattended began to come into the field, but most of them were not suitable for power generation.
5) The shortened work week in many industries increased operating problems and lowered the utilization of capital equipment.
6) A continual widening of the difference in average size of central power stations versus industrial power generators. For example, the Ford Motor Company 110 MW generators at River Rouge were seldom equalled by central station units at the time of their installation, but new utility units ten times as large are not uncommon now. Thus, economies of scale tilted in favor of central power stations.
7) An attitude on the part of industrial management that either said, in effect, "We're not in the power business," or that attempted to impose on power operations those policies relating to operation, maintenance, spare parts, etc., which apply to the major portion of the business. This latter can be frustrating to all concerned at best, and catastrophic at worst. For example, power plant maintenance usually comes in infrequent but expensive doses. In the desire to keep up production levels, the uninitiated manager may attempt arbitrary economies, which may appear successful because a power plant normally has substantial built-in reserves. Continuation of this practice, based upon its apparent success, can bring the manager to a point at which he suddenly finds himself without a functioning power plant. A large number of industrial plants have abruptly connected to the utility on a crash basis following the above scenario. The process has seldom been reversed.
8) Long-standing policies on the part of most privately owned electric utilities that have discouraged in every way possible the

generation of electricity by any other type of organization. Relevant here are rate schedules that favor large industrial users (whether justified or not by "cost of service"), and heavy demand charges (charges levied even if no power is used) that make it uneconomical to use the utility as a stand-by source backing up industrial power generation.

Industry has continued to generate steam because in most cases there was no other good alternative, but the factors just listed have led to a shift from on-site electric power generation to the present dependence on central station generation and distribution. The result has been convenience and flexibility at low apparent cost, and the potential economies of by-product power generation have been largely forgotten in many areas of industry.

Present Situation: According to the Stanford Research Institute (1), industry generated only 16.8% of its power needs in 1968. They estimated that if the trend downward continued, on-site generation in 1974 would amount to only about 14% of power consumption.

We have, then, in the U.S. two large relatively independent steam systems as illustrated in Figure 1.1A. A typical electric utility generates steam in a boiler, which drives a turbine-generator to produce electricity. In the process, the steam is condensed to hot water and the waste heat is rejected to the environment by means of a cooling tower or equivalent device. A typical industrial plant needing steam generates its own requirements in a boiler and consumes it on site, purchasing electric power from the utility except to the extent noted in the preceding paragraph. Only a very small percentage of steam is transferred from the utility side to the industry side.

This study explores two different modes of linkage between these more-or-less parallel energy systems. In Figure 1.1B, industrial steam is produced at a higher temperature and pressure than needed for process, and is brought down to the required state through a back-pressure turbine, generating electricity thereby before moving on to its intended end use. The by-product power resulting can be used locally or sold to the electric utility. In Figure 1.1C, some of the steam is extracted from the utility's turbine at the pressure required by an adjacent industry and supplied for its use, eliminating the need for a boiler in the industrial plant. In each of these cases, there are advantages and disadvantages which will be explored in more detail later in this report.

FIGURE 1.1: STEAM GENERATING SYSTEMS

A.

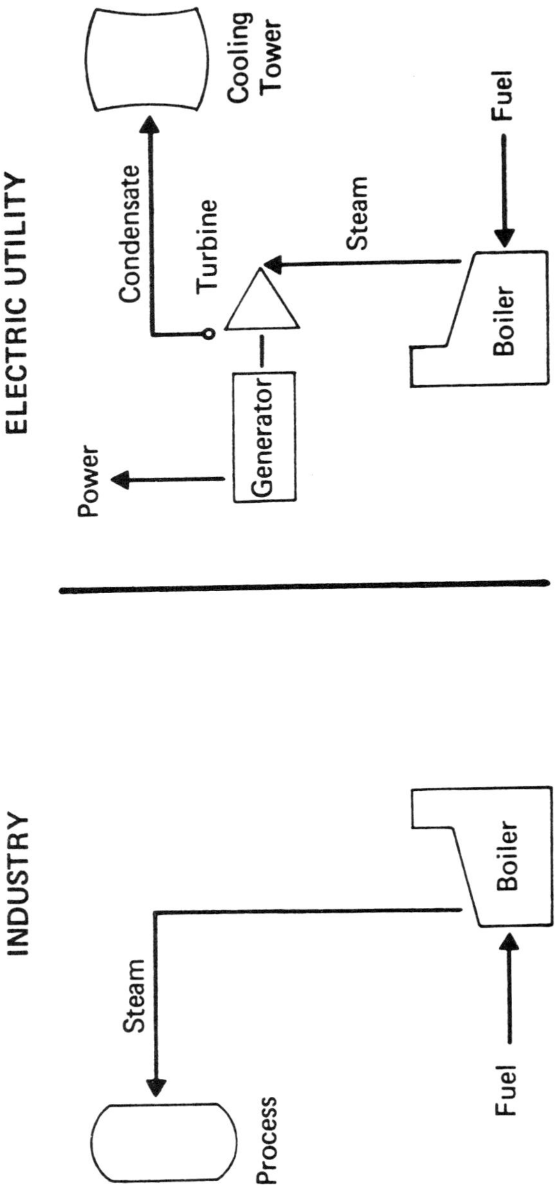

(continued)

FIGURE 1.1: (continued)

B.

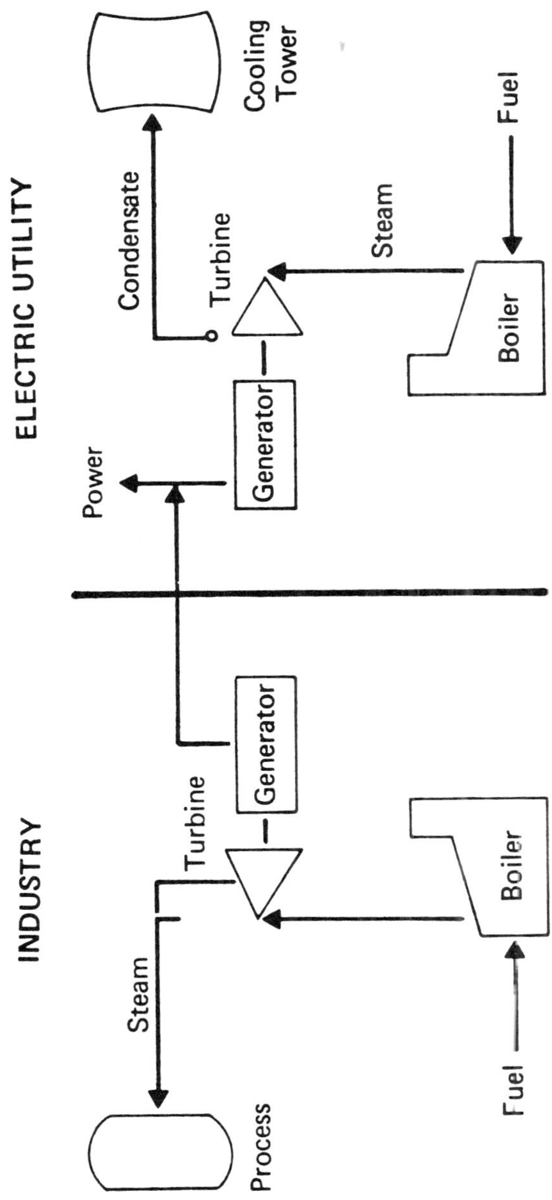

(continued)

FIGURE 1.1: (continued)

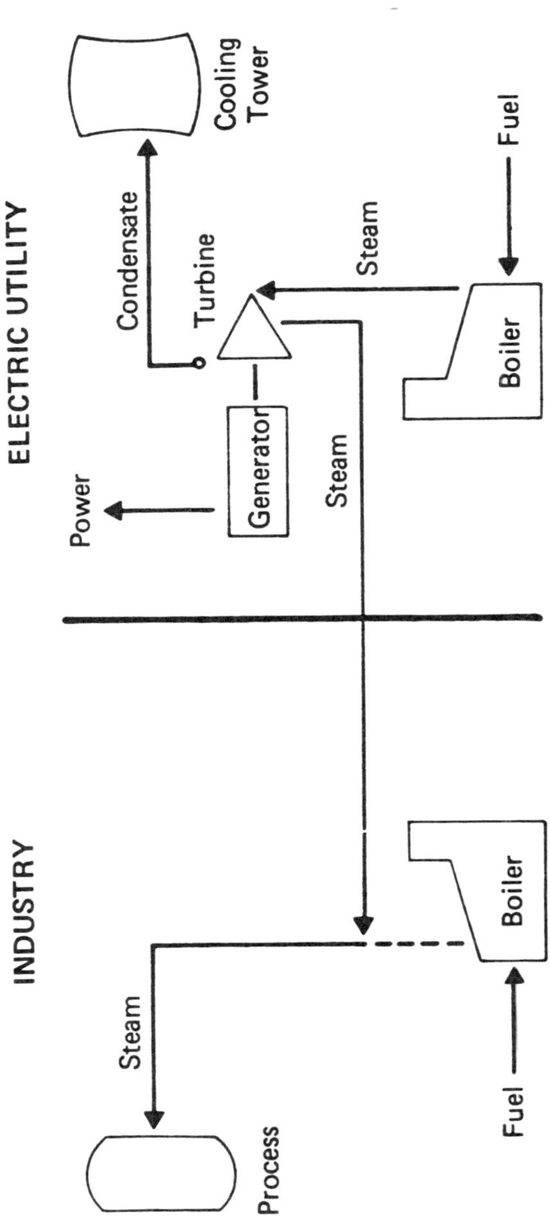

C.

The industrial portion of the energy demand is heavily concentrated in a relatively small number of segments of industry (2).

TABLE 1.1: INDUSTRIAL ENERGY CONSUMPTION PATTERN (INDUSTRIES)

		1968	1973
Primary Metals		21.2%	20.4%
Chemicals		19.8	21.4
Petroleum Refining		11.3	12.5
Food		5.3	5.3
Paper		5.2	4.9
Stone, Clay & Glass		4.9	4.6
	SUBTOTAL	67.7	69.1
All Other		32.3	30.9
	TOTAL	100.0	100.0

Industrial Boiler Trends: An additional factor of some importance in assessing the potential and best means of by-product power generation in industry is the characterization of existing boilers used for steam production.

Figure 1.2 shows what has happened over the past decade as coal has declined in popularity as a fuel in favor of oil and gas (9). At an even earlier period, coal was the major fuel for industrial boilers. Figure 1.3 summarizes this situation for industrial boilers sold during the five-year period 1968–1972, and emphasizes how much coal has really almost disappeared from the scene. It is outweighed almost five to one by the "other" fuels such as bagasse, black liquor, lignite, etc. In contrast with this, there has been a shift back to coal for electric utility boilers (21). During the period 1964 through 1969, coal-fired utility boilers accounted for 65% of new orders placed, falling to 49% in 1970 and 44% in 1971. This jumped to 75% in 1972, hit a new high of 90% in 1973, and was up to 98% for the first nine months of 1974.

Sometime around 1960, relatively inexpensive oil and gas fired designs were introduced, which accelerated the shift to oil and gas.

These were the "packaged boilers," pretty much assembled at the factory and shipped complete with fuel burning equipment, mechanical

FIGURE 1.2: PRIMARY FUEL NEW BOILER ORDERS

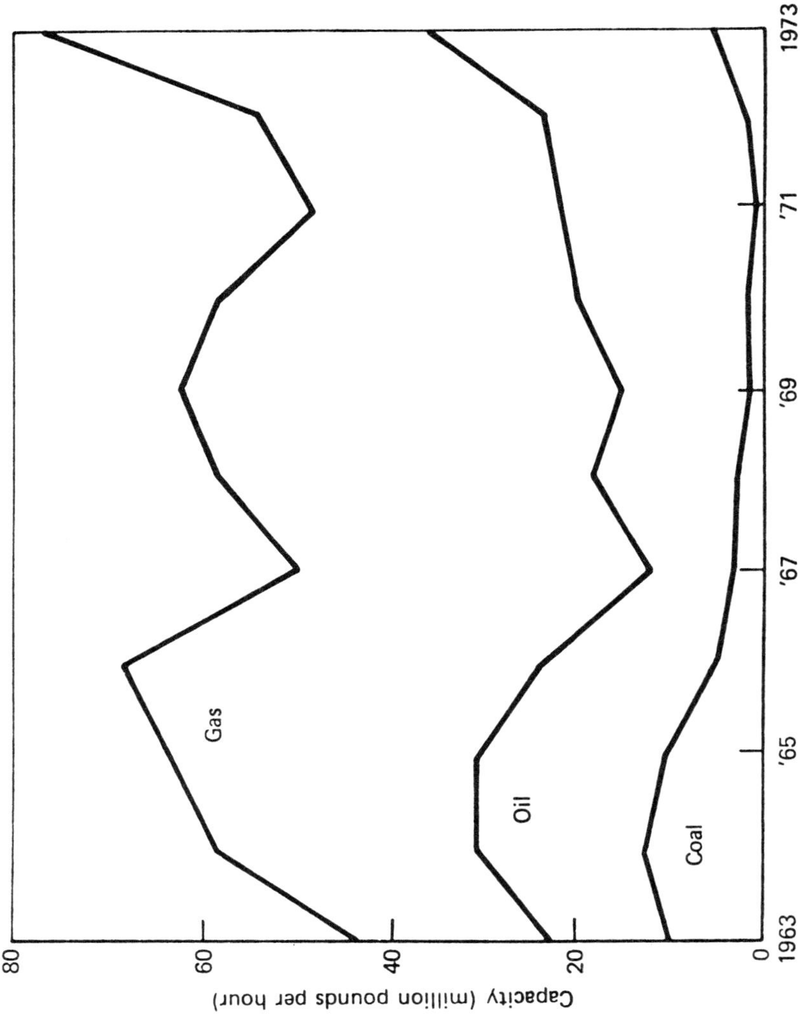

draft equipment, automatic controls, and accessories. Such boilers could be purchased on relatively short notice and even rented. They had the advantage of requiring a minimum of operating know-how and attention, and could be installed for 1/4 or 1/5 the cost of coal-fired, field-assembled boilers. They became exceedingly popular, and at the time, seemed almost an ideal answer to the need for industrial steam, especially in view of the increasing concern over environmental protection.

About 54% of industrial boiler capacity sold from 1968 through 1972 produced saturated steam (no superheat), and about 71% of the capacity came from packaged boilers.

FIGURE 1.3: INDUSTRIAL TYPE BOILER SALES 1968–1972

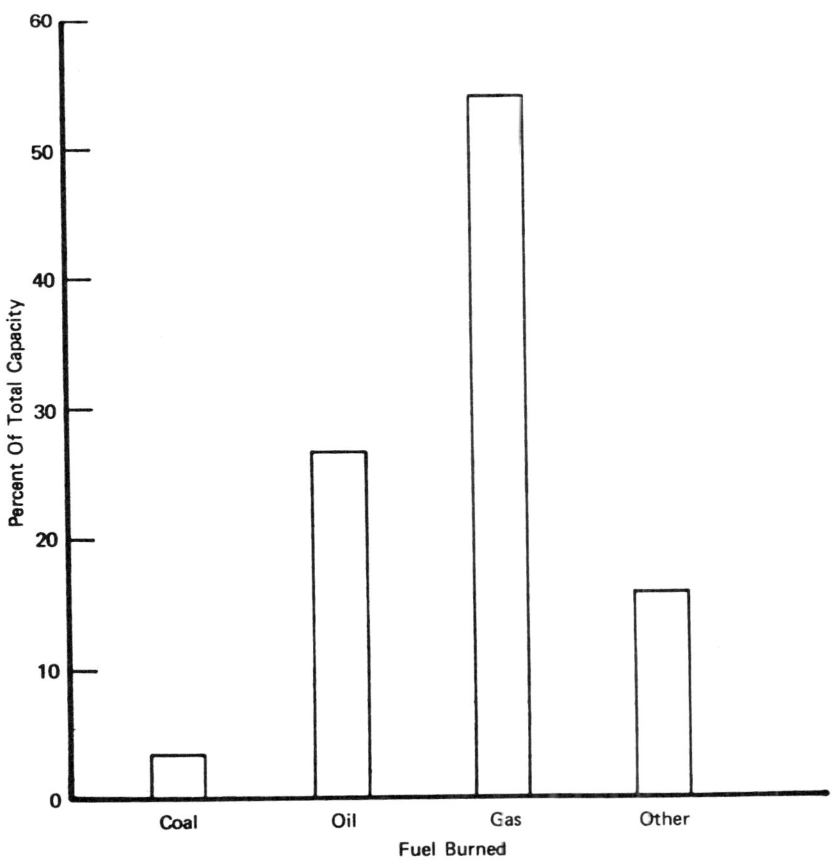

FIGURE 1.4: FUEL COST

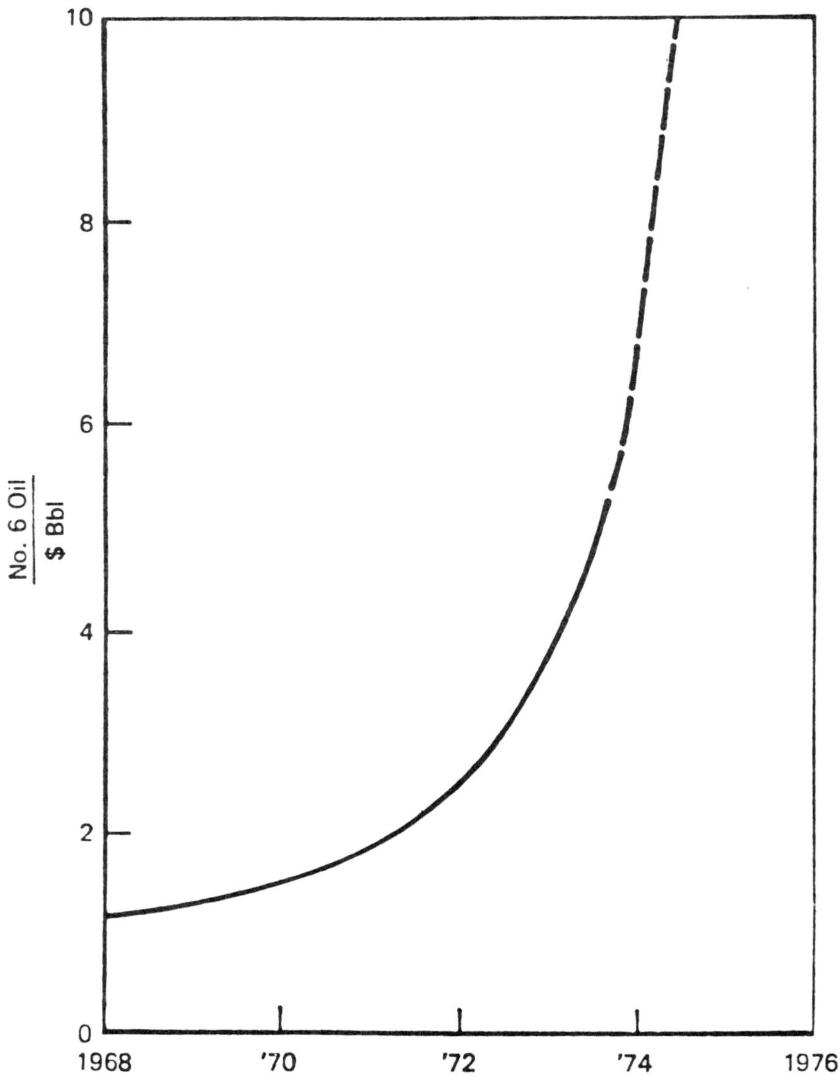

Maintenance can be a problem with packaged boilers. Small size and light weight are achieved by using small tubes of thin steel, closely spaced, which are in consequence vulnerable to feed water problems and exceedingly difficult to replace. Leaking tubes are usually just plugged off. If Venezuelan oil containing much vanadium is burned, boiler life can be very short, due to fireside corrosion of the tubes. In one such case, the boiler failed in a year and could not be repaired economically.

The design of package boilers was simplified to adapt to mass production methods. Most of them, for example, are without much in the way of heat traps. The operating efficiency is low, perhaps an average of 75% or less and seldom over 80%. This was not a prime consideration when fuel was cheap and plentiful, notably interstate natural gas, but the situation has changed markedly during the past few years, as may be seen in Figure 1.4 (9). The soaring cost of fuel has eaten up much of the capital advantage of the packaged boilers, and they are not suitable for burning coal or waste fuels, in general. Present and future supply problems with natural gas and oil add a further dimension to the diminishing attractiveness of packaged boilers.

The majority of existing packaged boilers are unsuitable for power generation, principally because of the pressure and temperature of the steam they generate, as well as the fuel problems just discussed. Nevertheless, they are present throughout the industry in large numbers and that presence must be a factor to be considered in this study.

By-Product Power Generation

General Aspects: By-product power is electricity generated from steam that is needed for another use and is suitable for that use after passing through the turbine. The power is thus considered as a by-product of steam production and use in developing the economics. This usually means employing an extraction or back-pressure turbine, and a variety of arrangements are possible (10).

In greatly simplified form that ignores machine-dependent energy losses, the basic idea that makes by-product power generation potentially attractive in comparison to central station power is illustrated in Figure 1.5. In the central power station case, energy must be added in the boiler to the feedwater in an amount sufficient to bring it up to point (A). However, the condensing turbines which drive the generators can only utilize the amount of energy between points (A) and (C) to

FIGURE 1.5: CENTRAL POWER vs BY-PRODUCT POWER GENERATION

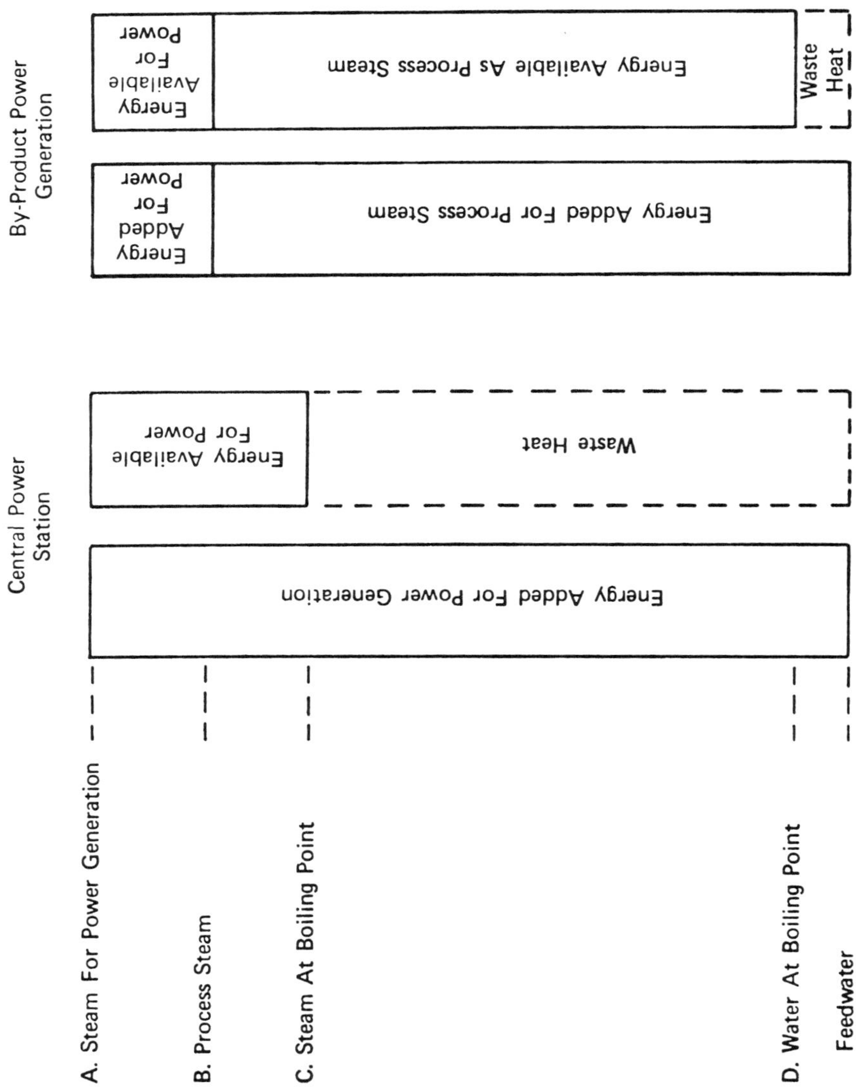

make electricity, and the large amount of energy from point (C) back down to the feedwater level must be rejected to the environment. In the by-product power case, the energy added from the feedwater level up to point (B) is already needed in order to generate the proper kind of steam for process and other industrial needs. On top of this an extra amount of energy is added to bring the steam up to point (A). The energy from (A) to (B) is now available for power generation, that from (B) down to (D) for process use as steam, and a relatively small amount is left as waste heat.

It should be apparent from this that by-product power has potentially several advantages over central station power. It consumes substantially less fuel per kWh of electricity produced, as has been recognized for many years (22). It creates much less of a problem with "thermal pollution" of the environment. It may or may not have a capital advantage, depending on the size and type of operation. These economies provide one of probably several reasons why some of the larger industrial steam users still generate power also.

In the more detailed analysis that follows, many industry averages and typical values of various quantities will be used, of necessity. At the same time, it is well to recognize that a specific situation may deviate from this picture in a number of respects. Industrial power rates from the local utility will vary over a considerable range, as will the terms for purchase (11). Fuel costs and fuel availability will be different. Plant investment figures will reflect varying labor rates and environmental restrictions. Other differences will include the pressure and temperature of steam required by industry, balance between industrial steam and electric loads, policies and financial position of the local utility, policies and practices of the Public Service Commission, etc. Thus it can be expected that what is economically viable at one specific location may not be so at another, and vice versa.

Ground Rules: Some of the principal basic assumptions made in the analysis to follow are:

1) Conditions were evaluated as they are projected to be in the year 1980. This was taken as the target year with explicit realization that full implementation by then might be unrealistic in view of probable lags in educational and decision processes, legislative changes, and equipment deliveries. Uncertainties multiply, however, as one goes further into the future.
2) All monetary figures were stated in 1974 dollars, unless otherwise identified.

3) The assumed rate of inflation was 10% in 1975, 5% per year until 1980, and 4% per year thereafter. The consumer price index was taken as indicator of inflation in analyzing historical data.
4) From Federal Power Commission data (12) it was estimated that power plant investment per unit of generating capacity has escalated 2.75% per year in excess of the rate of inflation. This was assumed to continue.
5) Interest rates for utility borrowings were assumed to be 4% per year in excess of the 5% per year mid-term rate inflation, i.e., 9% per year. This brings the utility industry average fixed charge rate from 15.02% of net utility plant per year in 1972 up to 16.87% in 1980, other factors remaining unchanged.
6) Extended lead times introduced by bottlenecks and shortages in fuel supply, steel supply, equipment manufacturing facilities, trained personnel, and the like were ignored. This could be a whole study in itself.
7) Except where stated otherwise, industrial steam and power plants were assumed to operate at full capacity 85% of the time. This was in contrast to a utility industry average of around 55%.
8) Fuel costs in 1980 for an average location are projected to be $1.25 per million Btu for high-sulfur coal, $0.30 per million Btu for nuclear fuel, $3.32 per million Btu for high-sulfur residual fuel oil, and $3.82 per million Btu for low-sulfur residual fuel oil.
9) Industrial power economics were based upon coal as boiler fuel, with investment figures containing allowances for particulate and sulfur dioxide control. It is not expected that natural gas will be available generally for this use in 1980, and the economics will greatly favor coal over oil in addition to oil availability problems (21).
10) By-product power was costed out on an incremental basis, i.e., the cost allocated to power was the cost of generating both steam and power minus the cost of producing only the steam required.
11) The measure used in evaluating economic viability from the industrial viewpoint was annual return on investment before tax. More sophisticated measures could be employed, but the broad nature of the study hardly justifies this.
12) For a basic operation like power generation, a 20% return on investment was considered satisfactory. In individual cases, this would vary, of course.
13) Industrial power plant configurations were based on practical considerations of design and operation as far as possible without becoming inappropriately detailed.

14) Equipment of standard size and characteristics were assumed.
15) Maximum power generation for a given process steam load was assumed.
16) To avoid undue proliferation of cases, all process steam was assumed to be 150 psig saturated.
17) Straight-line depreciation was used, with a plant life of 30 years.
18) Property taxes and other miscellaneous capital-related costs were assumed to be 2% of original investment per year.
19) Incremental working capital (mostly a small increase in fuel inventory) was ignored.

Utility System Power Stations

Basic Assumptions: In order to evaluate properly the economic desirability of industrial by-product power generation, a picture must be developed for the utility industry in 1980. This requires making a number of projections from various data and estimates currently available as will now be described.

The following table projects the mix of gross additions to the investor-owned utilities' generating capacity (13)(14)(15).

TABLE 1.2: MIX OF GROSS ADDITIONS TO GENERATING CAPACITY

YEAR OPERATIONAL	COAL	OIL	GAS	NUCLEAR	COMBINED CYCLE	PEAKING GAS TURBINES
1974	29.8%	18.9%	4.7%	31.7%	2.3%	12.6%
1975	40.2	17.9	0.9	25.9	2.5	12.6
1976	39.9	16.9	0	30.5	3.8	8.9
1977	47.0	17.3		19.6	5.5	10.6
1978	57.0	7.4		25.5	1.5	8.6
1979	58.7	0		30.3	2.0	9.0
1980	45.1			45.5	1.9	7.5
1981	28.5			59.8	2.8	8.9
1982	22.8			64.6	4.3	8.3
1983	22.6			61.0	7.3	9.1
1984	19.0			61.1	10.6	9.3
1985	14.4			61.9	14.2	9.5

It is assumed that additions to hydroelectric capacity will be quite small. These figures do not reflect the marked decline in nuclear plant orders recently (16). The percentages for nuclear additions are probably too high, but no attempt has been made to adjust them.

Fuel price forecasts can be found that range almost literally all over the map. Scenarios and projections were made in the study, as shown in Table 1.3. The only brief that can be held for these figures is that they are set down with fairly wide acquaintance with other forecasts and trends of thought, and out of considerable experience in following and forecasting fuel markets. No other claims or apologies are made.

TABLE 1.3: PROJECTED FUEL PRICES ($/MILLION Btu)

YEAR	HI-S COAL	NUCLEAR	AVERAGE UTILITY GAS	CRUDE OIL & HI-S RESID.	LOW-S RESID.	#2 DISTILLATE
1974	0.58	0.22	0.48	2.00	2.30	2.50
1975	0.69	0.23	0.60	2.26	2.60	2.83
1976	0.80	0.25	0.74	2.44	2.81	3.05
1977	0.91	0.26	0.92	2.64	3.03	3.30
1978	1.02	0.27	1.11	2.85	3.27	3.56
1979	1.14	0.29		3.07	3.54	3.84
1980	1.25	0.30		3.32	3.82	4.15
1981	1.37	0.32		3.52	4.05	4.40
1982	1.50	0.33		3.73	4.29	4.66
1983	1.62	0.35		3.95	4.55	4.94
1984	1.74	0.37		4.19	4.82	5.24
1985	1.87	0.39		4.44	5.11	5.55

It is interesting to note the electric utility fuel costs reported for April 1974 (18): coal $0.64 per million Btu, oil $1,872 per million Btu, and gas $0.436 per million Btu. During the year from April 1973 to April 1974, coal burned by utilities went up 62% in cost, oil up 169%, gas up 31%, and weighted average of all three fuels up 85%. During the same period the consumer price index only went up 10%, and the wholesale price index up 17%.

Utility average heat rates are assumed to continue at 10,300 Btu per kWh for coal, oil, gas and nuclear plants, 9,200 for combined cycle plants, and 15,000 for peaking units.

For those used to thinking in terms of physical quantities rather than Btus, a barrel of crude oil contains about 5,800 Btu and of residual fuel oil, about 6,100,000 Btu. A ton of coal might average 22,000,000 Btu, but will vary quite a bit, while a cubic foot of gas may average 1,025 Btu.

Operating and maintenance costs of individual power stations are derived from a chart in Reference 19. For the utility industry as a whole, operating and maintenance costs are determined as a projected percentage of utility plant investment.

System Averages in 1980: A picture of the electric utility industry in

1980 was developed from the information in the preceding section and building upon the composite data for the industry as a whole as reported by the Federal Power Commission (20). The main purpose of this exercise was to estimate an average system rate for electric power to industrial customers in 1980; the details of deriving it are less important. The extent and character of additions to generating capacity may differ considerably from the course assumed in this study, but as can be seen, this would not have a large impact on the 1980 industrial rate (see Table 1.4).

TABLE 1.4: ELECTRIC UTILITY FUEL COSTS IN 1980

	BILLION KWH	$ MILLIONS
Coal	1,113	14,327
High-S Resid.	605	20,676
Low-S Resid.	110	5,442
Nuclear	551	1,704
Hydro.	75	-
	2,454	42,149

The industrial electric load is projected to be 892.9 billion kWh in 1980, to which 119,917 MW of generating capacity would be dedicated (at 85% load factor). Costs in 1980 might then be allocated as in the next table.

TABLE 1.5: ELECTRIC UTILITY COSTS IN 1980

	INDUSTRIAL	RESIDENTIAL, COMMERCIAL & OTHER
Billion Kwh	892.9	1,561.4
Fixed Charges		
Production & Transmission	$6,255 MM	$18,151 MM
Distribution	-	11,795
Operating & Maintenance	4,065	11,795
Overhead	2,257	3,947
Fuel	15,334	26,815
TOTAL COSTS	27,911	72,503
Mills per Kwh	31.26	46.43

Using the same set of figures, it is estimated that the 1980 industrial rate of 31.26 mills per kWh might be split into an average demand charge of $3.21 per month per kW and an energy charge of 26.09 mills per kWh (based also on typical industrial rates shown in Reference 11).

Many orders for new power plants have been either cancelled or deferred for substantial periods of time. Orders for nuclear power plants have been particularly hard hit (16). To gain some idea of the sensitivity of the 1980 system rates to the additions to generating capacity between now and then, an industrial rate was calculated for the hypothetical case of no additions after 1972. By 1980, the industrial rate would then be 31.87 mills per kWh, not significantly different from the rate in Table 1.5.

New Power Stations in 1980: In order to assess the economic viability of industrial generation of power for sale it is necessary to have a value for sale (presumably to a utility). Assuming equal reliability, the limiting value for sale will be the value of power at the bus bar for the utility's best alternative, which is taken to be a new central power station of average size planned to be operational in 1980 (13)(14). Since most new plants are either nuclear or coal-fired, both will be considered. Investment figures were developed from the ORCOST program (17).

TABLE 1.6: NEW CENTRAL POWER STATIONS IN 1980

	COAL-FIRED	NUCLEAR
Average Size Planned (MW)	662	1,092
Investment ($M)	352,700	601,300
Load Factor (%)	70	82.1
Power Generated (MM Kwh/Yr)	4,059.4	7,853.6
Assumed Heat Rate (Btu/Kwh)	9,000	10,300
Annual Costs ($M)		
Fixed Charges	59,260	101,020
Operating & Maintenance	3,470	5,000
Fuel	45,670	24,270
TOTAL	108,400	130,290
Bus Bar Power Value (Mills/Kwh)	26.70	16.59

If the power generated by industry is sold as peaking power instead of firm power, the limiting value for sale would be the bus bar value of power for new utility peaking units planned to be operational in 1980. Arbitrarily assuming such a unit to be 50 MW in size, costs would be as given in Table 1.7 for two fuel choices.

TABLE 1.7: NEW PEAKING UNITS IN 1980

	LOW-S RESID.	#2 DISTILLATE
Investment ($M)	8,880	7,992
Load Factor (%)	17.7	17.7
Power Generated (MM Kwh/Yr)	77.526	77.526
Assumed Heat Rate (Btu/Kwh)	15,000	15,000
Annual Costs ($M)		
Fixed Charges	1,492	1,343
Operating & Maintenance	300	270
Fuel	4,442	4,826
TOTAL	6,234	6,439
Bus Bar Power Value (Mills/Kwh)	80.41	83.06

General Description of Industrial Power Plants

It was pointed out previously that many, if not most, of the boilers presently used in industry only for steam generation are unsuitable for by-product power generation. New installations will be required, and the first basic assumption is that they will be coal-fired Table 1.8 presents a comparison of two ways of generating 400,000 pounds per hour of 150 psi process steam in 1980.

TABLE 1.8: ALTERNATIVE STEAM GENERATORS IN 1980

	PACKAGE BOILER	FIELD-ERECTED BOILER
Fuel	Oil	Coal
Useful Life (years)	15	30
Fuel Efficiency (%)	80	88
Investment ($M)	2,585	10,339
Annual Costs ($M)		
Depreciation & Misc.	224	551
Operating & Maintenance	100	377
Fuel	14,843	4,415
TOTAL	15,167	5,343

The reduction in annual costs in going from an oil-fired package boiler to a coal-fired field-erected boiler is sufficient to yield a 127% return (before tax) on the additional investment required. Calculations based on historical order rates suggest that the average age of the industrial package population is between 5 and 7 years. If a company owned a package boiler 6 years old in 1980, it could still achieve a 93% return (before tax) on the investment in a new coal-fired boiler to replace the oil-fired one. From this point on, therefore, we will be assuming that coal will be used for industrial steam generation, with or without by-product power production. At the same time, the possibility is recognized that small package boilers (say, less than 60,000 pounds per hour) might still be desirable for unattended operation.

Figure 1.6 illustrates a simple back-pressure turbine system that might be used for industrial by-product power generation, utilizing steam from a single coal-fired boiler, and possibly operated in parallel with the power system. Turbine driven boiler auxiliaries would permit starts with the power grid dead. An average start-up time of 2 hours could be expected from cold boilers, or 30 minutes from banked fires and 100 psi or more boiler pressure. The heat traps would consist of a small economizer and air preheater to supply air to the burners at 300°F. Firing could be with pulverized coal or a travelling grate spreader stoker, depending on size and a number of other considerations. The boiler would be provided with a good mechanical dust collector in order to trap cinders for refiring, followed by a single electrostatic precipitator or scrubber and sulfur removal equipment.

Two stages of feedwater heating are employed in order to maximize power production. The deaerating heater is supplied with exhaust steam from the boiler feed pump turbine and the high pressure feedwater heater uses steam from an intermediate extraction point in the main turbine. The turbine-generator is a standard noncondensing unit such as is presently available from several manufacturers. The entire unit could be automated so that one operator on each shift could start, stop, and perform all functions including coal handling and feedwater, condensate, and boiler water control.

Given a required quantity and pressure of process steam, an almost infinite number of different systems like that in Figure 1.6 could be postulated. In order to reduce the problem to manageable size, without at the same time oversimplifying it, it will be assumed that 150 psi saturated steam is typical of a majority of large-volume process requirements, and attention will be confined to that, and also to the more common "standard" sizes of turbine-generators and boiler temperatures and pressures.

A system like that of Figure 1.6 has the advantages of simplicity and efficiency, but lacks somewhat on the side of reliability and flexibility although it might serve for parallel operation with a utility system which provides back-up. With stand-by and demand charges as they are specified in typical utility rate structures, however, this could be a costly pattern of operation. If equipment is maintained in reasonable condi-

FIGURE 1.6: SIMPLE BACK-PRESSURE TURBINE SYSTEM

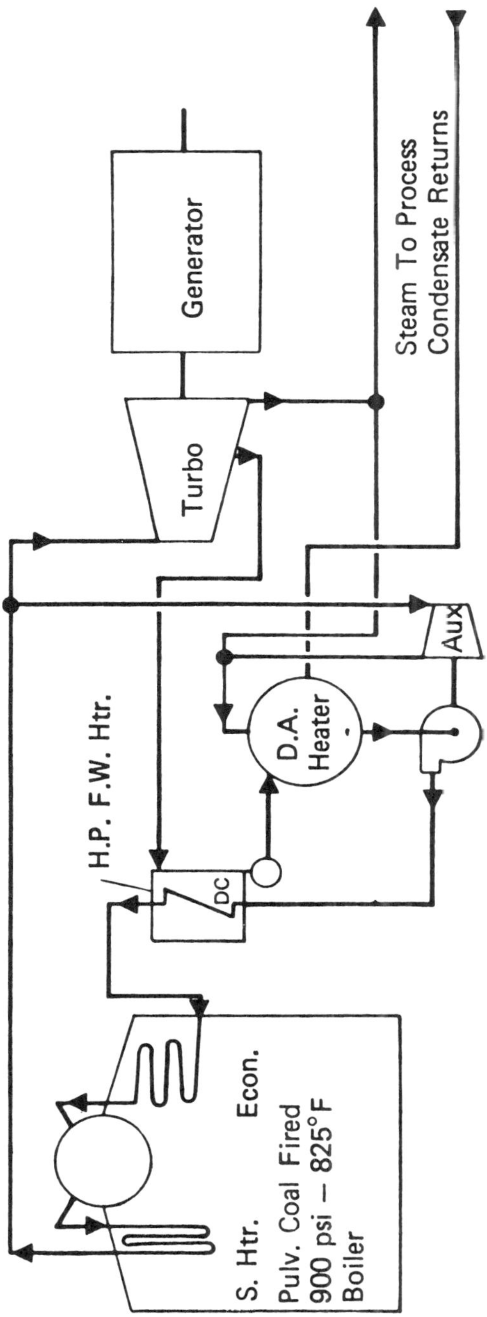

tion, most of the unscheduled outages will result from troubles with the boiler rather than the turbine-generator. An alternative system with improved reliability, as illustrated in Figure 1.7, was examined in some detail. Its essential features are the use of conventional low-pressure package boilers (which might already be in place) as back-up for the steam supply, and a mixed pressure turbine with a condensing exhaust.

The turbine is normally driven by steam from the high pressure coal-fired boiler, although it may also induct low pressure steam from the oil-fired package boilers during periods of heavy electric load. During lighter load periods it can also extract low pressure steam for process, allowing the package boilers to be shut down in favor of the much higher efficiency of the high pressure boiler. The turbine could be operated at reduced load by cooling the high pressure blading with low pressure steam and using sufficient low pressure steam in the low pressure nozzle to carry the required electric load. Thus, during main boiler outages, the package boilers could supply the process steam load and operate the turbine condensing, to supply part or all of the electric load. Package boilers without superheaters, which means most of them, can be very quick on emergency load pick-up, since there is not the usual danger of overheating the superheater. When the peak loads or emergencies are of relatively short duration, the moisture in the exhaust steam is of small concern. All of these features make this system more flexible and reliable than the simpler one of Figure 1.6, and it will be used as the main basis for a look at costs and economics.

Heat Balances

Heat balances were determined for a number of different systems and sizes of operation. Steam and water flows of a simple regenerative system like that of Figure 1.6 are shown in Figure 1.8, the upper half representing the generation of power and steam, and the lower half representing the generation of steam only.

For steam entering a turbine at 900 psi and 825°F and leaving at 150 psi, the appropriate tables (23) give a theoretical steam rate of 18.18 pounds per kWh. To convert this to actual steam rate, a cut and fit procedure involving turbine-generator efficiency was used (24), resulting in an overall engine efficiency of 81% and an actual steam rate of 22.44 pounds per kWh. Power needed for electrical auxiliaries was estimated to be 1,000 kW so that gross power generated was 21,000 kW. The

FIGURE 1.7: STEAM/POWER—AUTOMATIC EXTRACTION—MIXED PRESSURE CONDENSING TURBINE

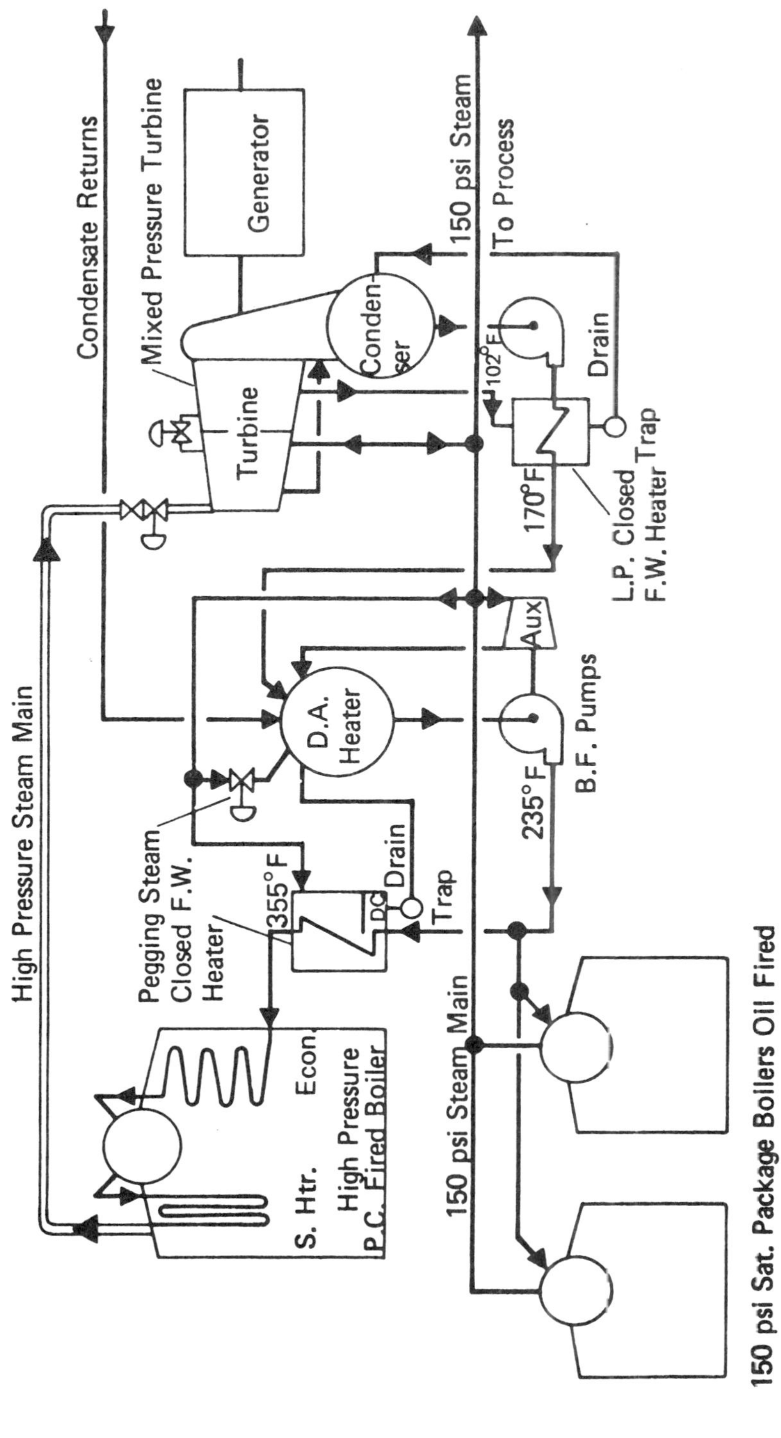

throttle flow to the turbine was then 22.44 x 21,000 = 471,240 pounds per hour. With a 5% blowdown, the feedwater flow to the boiler is 494,802 pounds per hour. The temperature of the feedwater will be just a few degrees below the saturation temperature of the 150 psi exhaust steam going to the high pressure feedwater heater, or say 355°F (25). The water going to the H.P.F. heater from the deaerating heater will be just a few degrees below the saturation temperature for 10 psig, at which the deaerator operates, or say 235°F.

The condensate going from the steam side of the H.P.F. heater to the deaerator cannot be more than 15°F hotter than the deaerator, or say 250°F. Making a heat balance across the H.P.F. heater, then, gives a steam flow to it of 58,689 pounds per hour. Making heat and water balances across the deaerating heater gives a condensate plus make-up flow of 414,701 pounds per hour and a 10 psig steam flow of 21,412 pounds per hour. 100% return of condensate was assumed and the make-up water flow is 23,562 pounds per hour so the flow of 150 psi steam to process must be 391,139 lb/hr. The actual steam rate of the boiler feed pump turbine is 24.32 pounds per horsepower-hour, and at 800 horsepower, this requires 19,456 lb/hr which exhausts into the deaerator. The balance of steam to the deaerator, or 1,956 lb/hr is pegging steam to maintain pressure in the deaerator.

The heat in the blow-down is pretty much recovered in flash tanks, so that the input heat required from burning the fuel is essentially that needed to convert 471,240 lb/hr of water at 355°F into steam at 150 psi and 825°F. This amounts to 578,945,000 Btu per hour at a boiler efficiency of 88%. In order to determine how much of this energy was needed for power generation, a heat balance must be determined for producing the equivalent amount of process steam only, as shown in Figure 1.8B.

Since the 391,139 pounds per hour of process steam in Figure 1.8A is superheated and that in Figure 1.8B is saturated, it is necessary to convert to an energy-equivalent amount, which turns out to be 411,172 pounds per hour. Working backward from this, the total steam flow from the boiler is 432,626 pounds per hour and 487,442,000 Btu per hour are required to generate this from water at 235°F, again at an 88% boiler efficiency.

The difference between the two fuel consumption rates is the energy required to generate the 20,000 kW of by-product power, or 4,575 Btu per kWh. If to the 20,000 kW the 900 kW required by electrical auxiliaries

FIGURE 1.8: SIMPLE REGENERATIVE SYSTEM

A. Generation of Power and Steam

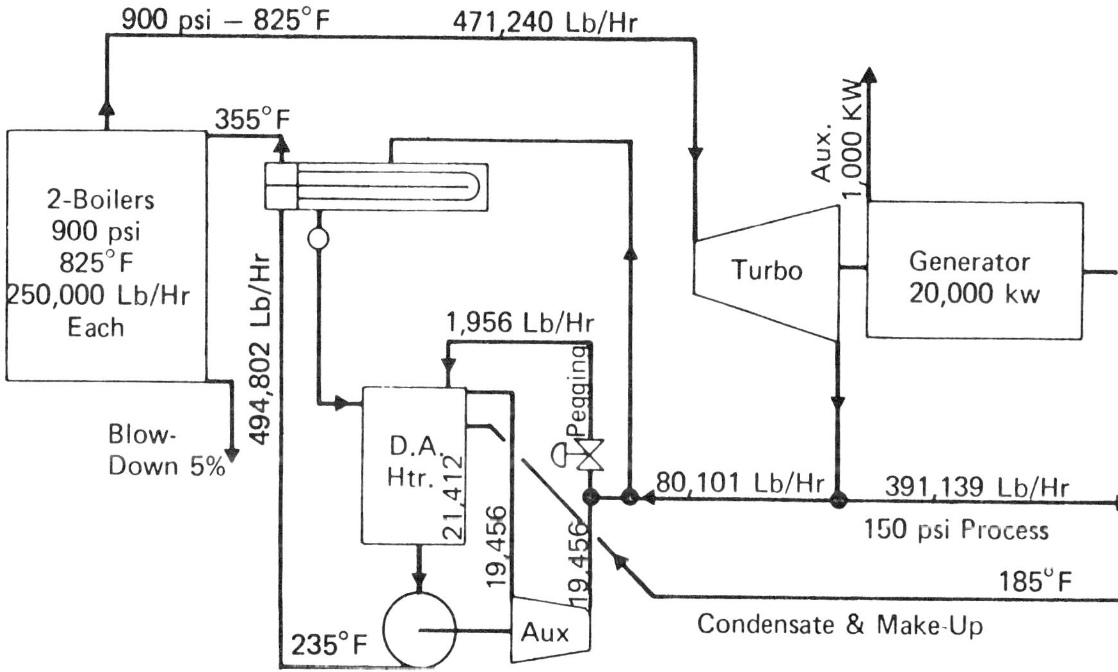

(continued)

FIGURE 1.8: (continued)

B. Generation of Steam Only

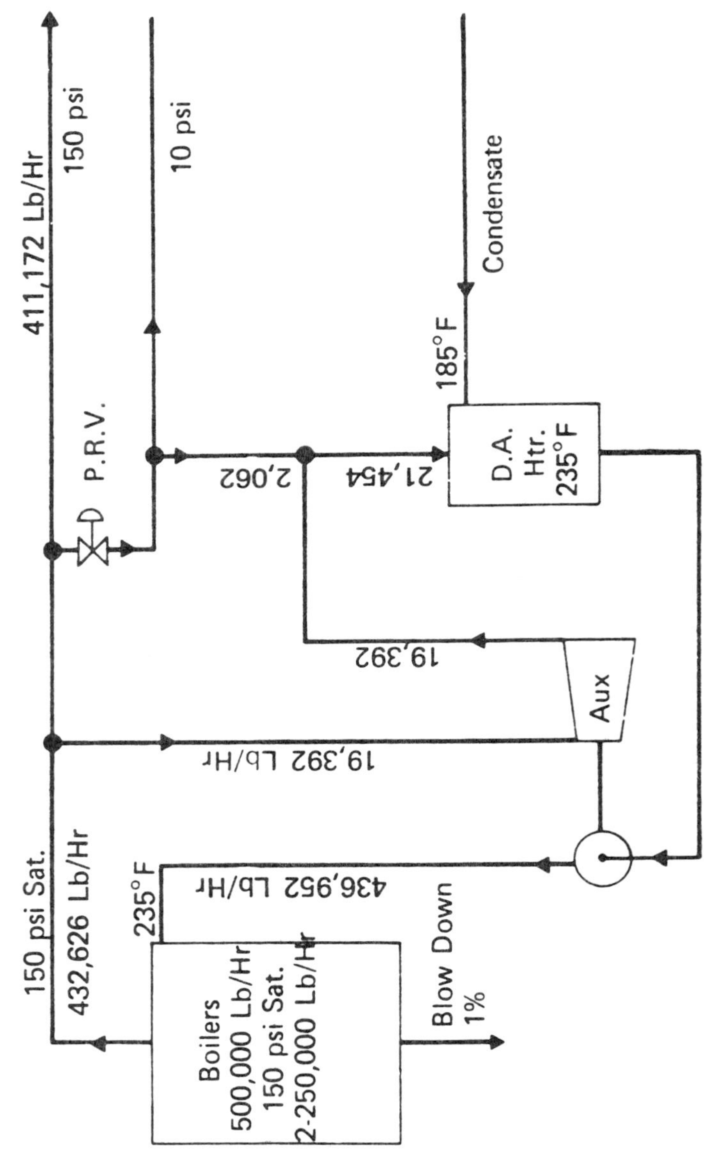

in producing steam alone is added, the heat rate becomes 4,380 Btu per kWh. A number of such power cycles of different capacities were studied and an average heat rate of 4,550 Btu per kWh obtained.

It should be noted that, for any specific power generation system, even simple ones, the matter of heat balances is a relatively complicated process.

Disposal of By-Product Power

Industry Patterns: The first obvious use for any additional generation of power by industry would be to replace power purchased from the local utility. The extent to which this is possible varies widely from plant to plant and even between general industrial categories.

Studies showed that industry consumed 82% more electric power than it could have generated from its steam base in 1967, and only 30% of that steam base was used for power generation.

With certain exceptions, then, it would appear that most industries could consume locally all of the by-product power they may generate, and that this will not meet all their needs in many cases. It can be economically feasible to meet this extra need in many cases by increasing the amount of condensing power generated in a system such as that illustrated in Figure 1.7.

Similarly, it was found that it can be economically feasible to generate still more condensing power and sell the surplus to an electric utility for distribution and resale.

Wheeling Power: There will undoubtedly be occasions when the local electric utility is not interested in purchasing surplus power from an industrial plant, indeed, this has been the norm in the past. Such a situation can arise because the utility feels that it has the capacity to meet projected loads, because it is concerned about adequate controls, because it wishes to discourage industrial generation, or just because that has been its traditional policy. In some such cases, however, a nearby utility might have need for the surplus power, and this raises the question of wheeling the power over the local utility's transmission lines to the neighboring utility's system.

The term "wheeling" is used to denote the transmission of electric power from one producer to a second party over the transmission lines of a third party. It represents a service performed by a utility to one or more other parties. The most common application of this technique involves the distribution of power from the federal power generation facilities to various utilities (private, municipal, public) which market the power. The development of other large generating facilities (such as pumped storage, large nuclear plants, etc.) has encouraged the wheeling of power. Furthermore, the joint ownership of large, high capital plants by a number of utilities also serves to encourage the wheeling of power back to the owners for marketing.

Many considerations would be involved in wheeling operations between industry and neighboring utilities, but in general figures show that the number of wheeling operations has grown considerably in the past few years.

The largest single utility user of wheeling operations has been the Bonneville Power Authority which serves as a marketing agent for a number of federal power generating agencies. This organization accounts for more than half of the wheeling done by public utilities. It also serves as a "broker" in arranging sources of power on spot or short-term basis for various utilities. In this case, a customer approaches Bonneville with a power requirement, Bonneville locates a source of power to fill the need and wheels the power from source to customer for a fee.

There have been no examples located in which power is wheeled from or to an industrial organization on a long-term basis. Several short-term agreements whereby industrial power has been wheeled by Bonneville to a utility have been observed. One example of this involves the industry operation of a Public Utility District power plant using scrap wood as a fuel. Power is sold to a utility via wheeling agreement with Bonneville. This arrangement serves as a short-term power supply (1-2 year contracts). In this case, by-product steam is also sold to industry.

Service charges for wheeling of power can be based on either of two criteria:

a. Peak transmission capacity used; or
b. Energy transmitted times distance moved (sometimes including a percentage energy loss).

The first criteria is generally aimed at the recovery of the capital plus operating and maintenance costs associated with the installation of extra transmission capacity. The charges, then, are independent of the total quantity of energy transmitted but are affected by the maximum demand rate involved (as determined over some relatively short 15 minute to 1 hour time intervals). Some components in the transmission system would be independent of wheeling distance (transformers, switch gear, etc.) whereas other components would depend upon the distance covered (transmission lines, towers, etc.). In the ultimate case, the wheeling charge would be made up by summing the individual component charges as is shown in Table 1.9. Such a detailed analysis is practical only in a system wherein the transmission operation can be broken down in detail. An example of the annual cost using Table 1.9 might be the determination of the wheeling charge to an industrial load at 115 kV wherein the power is transmitted 200 miles to a main grid substation.

Annual cost per kW capacity:

Line Terminal	$0.15
Transmission Distance 0.014 x 200	2.80
Switching, Transformation & Backup	0.60
Line Terminal	0.15
Transformation	1.00
Terminal	0.14
	$4.84/kW per year

If this hypothetical customer has a use factor of say, 90%, the annual cost when distributed over the energy used becomes:

$$\$4.84/365 \times 24 \times 0.9 = 0.614 \text{ mills per kWh}$$

On the other hand, if the use factor is only 20%, the cost becomes 2.7 mills per kWh and wheeling is much less economical.

It should be noted that the figures shown in Table 1.9 are based on a 6⅛% ultimate interest on plant additions which is representative of Bonneville expectations. Wheeling charges when calculated by a private utility would be somewhat higher to reflect the higher cost of capital. In fact, one such contractual agreement involving a private utility shows an annual charge of $5.65 (as compared to $4.84 shown above) subject to escalation and power factor penalties. Direct comparison of the two numbers is not possible since the distance over which transmission is contemplated is not necessarily the same. However, the effect of increased capital cost is apparent.

TABLE 1.9: BONNEVILLE POWER ADMINISTRATION—STANDARD WHEELING FORMULA
(charges per kilowatt-year)

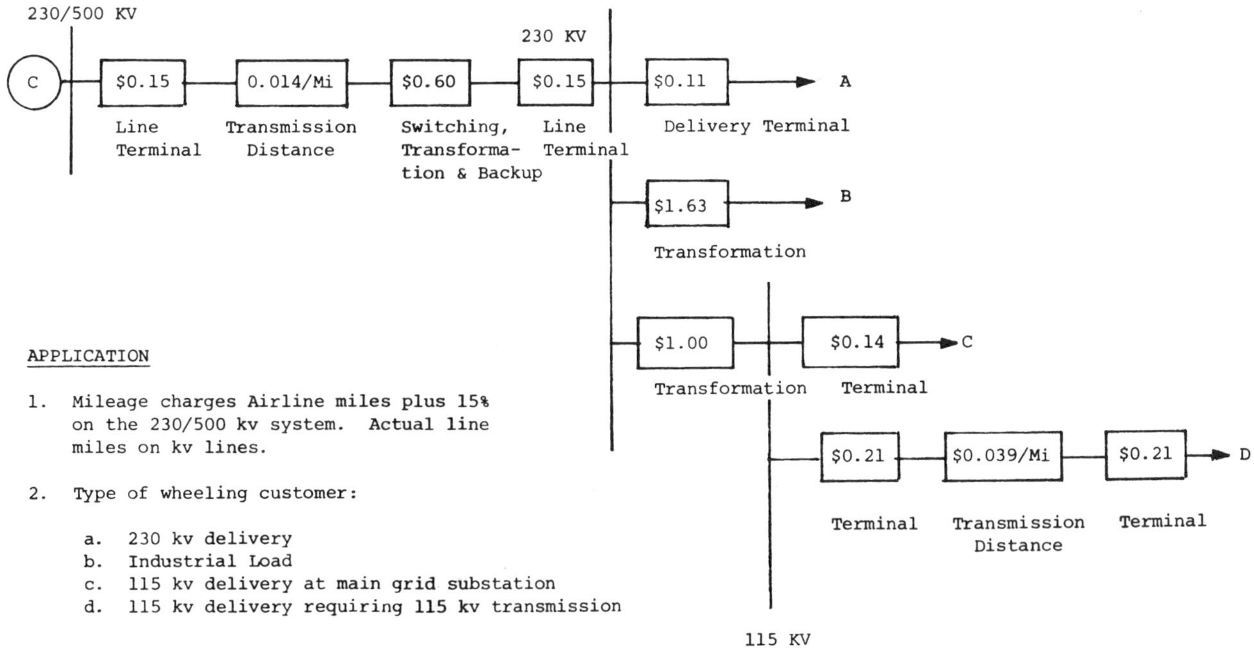

APPLICATION

1. Mileage charges Airline miles plus 15% on the 230/500 kv system. Actual line miles on kv lines.

2. Type of wheeling customer:

   a. 230 kv delivery
   b. Industrial Load
   c. 115 kv delivery at main grid substation
   d. 115 kv delivery requiring 115 kv transmission

Systems Considerations: It is expected that there will be little by way of intrinsic electrical systems problems caused by dispersed industrial generation. On the whole, there are net benefits from a systems standpoint of having a larger number of smaller power sources distributed throughout a system. Such problems as may arise are related more to operating procedures or institutional barriers than to any inherent technical problems.

Careful planning is required to integrate the industrial generation into the utility system to avoid operating problems and to maximize the benefits of efficient and low-cost generation sources. The principal requirement is the predictability of the industrial generation to allow rational planning and operation of the utility system.

Large quantities of remote industrial generation would require especially careful planning for their economic use, but present no new technical problems. No special problems of transient stability are seen as arising from the utilization of industrial generation.

One must distinguish between the result of merely dumping new sources of generation on an existing system and a comprehensive generation-transmission plan which includes such sources. In the former case, operational difficulties may well arise. In the latter, the transmission system can be designed to accommodate the distributed sources for satisfactory operation. If the system is designed with provision for industrial generation sources, the total transmission requirements will almost certainly be less than they would be if all power were supplied by conventional large central stations. Average distances between sources and loads will be reduced. This means shorter transmission lines of lower voltage.

There is currently great interest in both energy storage devices and small generation devices such as fuel cells which can be distributed throughout a system at substations. Both are desirable because they reduce transmission as well as generation capacity requirements. Properly integrated industrial generation can have the same effect.

This rather favorable picture of the effects of distributed industrial generation is based on the tacit assumption that the generating units are controlled for the benefit of the electric utility and that their availability is comparable to the utility's own units.

A completely arbitrary operating policy—with power supplied from the industrial generation according to the operation of the industrial process or even the whims of its operators—could be accommodated and would not necessarily be detrimental to the operation of the electric utility system. Under these conditions, however, it is unlikely that any utility would give a capacity credit for the generation and would, at most, pay for any available power at low "dump" energy rates. Their argument is that if the industrial generation is not as reliable as their own, or if they cannot control or at least schedule its output, then they would have to provide (duplicate) capacity. Thus potential capital savings disappear and capacity credit cannot be given. This applies to transmission capacity as well as generation.

The price paid for dump energy is only a little above incremental operating costs. Even with the very favorable economics of combined power generation and process heating this will not be sufficient to recover the capital costs of excess generation. It thus becomes uneconomical for an industrial organization to produce excess electrical power under these conditions.

Power Economics

By-Product Power: We will consider first the generation of by-product power, i.e., power generation capability that just matches the process steam load. This is calculated to be about one kW for every 20.4 pounds per hour of process steam, under the conditions that have been assumed. A (nominally) 20 MW mixed pressure turbine system backed up by package boilers will be used as an example. An overall operating factor of 85% is used, with the oil-fired package boilers being required 12 days per year. The package boilers operate at 300 psi and cost one-fourth as much as equivalent field-erected boilers. Results are shown in Table 1.10.

If the power is all used locally by the industry, its value is 31.26 per kWh. The before-tax return on the incremental investment in power is then 17.6% per year.

There may well turn out to be many situations in which oil-fired package boilers are available at costs below those considered normal, either because of reduced demand due to fuel supply and cost problems or because a partially-depreciated one is already on hand. A repeat of the preceding calculations with the package boiler investment reduced 50% results in by-product power at 15.94 mills per kWh and a before tax return on investment of 22.5%.

TABLE 1.10: COSTS FOR 20 MW BY-PRODUCT POWER SYSTEM IN 1980

		STEAM ONLY	STEAM & POWER	ALLOCATED TO POWER
Power generated & Saved (M Kwh)		-	160,626	160,626
Process Steam (Pounds/Hr)		440,186	414,720*	-
Investment: ($M)	H.P. Boiler	-	15,592	
	L.P. Boiler	11,704	4,702	
	Power	-	4,700	
	TOTAL	11,704	24,994	13,290
Annual Costs: ($M)	Depreciation	390	833	
	Optg. & Maint.	375	570	
	Misc.	234	500	
	Fuel	4,830	6,612	
	TOTAL	5,829	8,515	2,686
Power Cost (Mills/Kwh)		-	-	16.72

* This has the same heating value as the "steam only" figure.

An extension of this calculation to other sizes of operation is shown in the following table.

TABLE 1.11: RETURN ON INVESTMENT IN BY-PRODUCT POWER

NOMINAL	PACKAGE BOILERS AT NORMAL COST		PACKAGE BOILERS AT 50% COST	
GENERATOR SIZE	MILLS/KWH	% ROI (BT)	MILLS/KWH	% ROI (BT)
5 Mw	20.65	9.4	19.67	12.3
10	17.73	14.1	16.86	18.0
20	16.72	17.6	15.94	22.5
30	16.21	19.8	15.48	25.4
40	15.88	21.6	15.18	27.6
60	15.46	24.1	14.80	31.0
100	14.99	27.6	14.38	35.5

Using a return on investment (ROI) of 20% as criterion of economic viability, and taking into account probable uncertainties in package boiler costs, it is concluded that by-product power installations above approximately 400,000 pounds per hour of process steam or 20 MW of by-product power generation will be viable. This is, of course, a rough-cut criterion based on averages and conditions believed to be typical; there will be exceptions for specific cases above and below this level and individual situations must be evaluated on their own merits.

Incremental Condensing Power: It was noted previously that much of industry uses more electric power than it could generate from its steam base at the rate of one kWh for every 20.4 pounds of steam. The mixed pressure turbine has a condensing end so it seemed reasonable to consider the economics of generating the additional power by increasing the flow through the condenser. One could not expect a better heat rate than those normally found for condensing power, but since associated investment and operating costs are incremental there could be some economic advantage. An example calculation is shown in Table 1.12.

TABLE 1.12: INCREMENTAL CONDENSING POWER FOR USE

	20 MW BY-PRODUCT POWER	20MW BY-PRODUCT PLUS 10 MW CONDENSING	ALLOCATED TO 10 MW CONDENSING POWER
Power Gen. & Saved (M Kwh)	160,626	235,086	74,460
Investment ($M)	24,994	28,866	3,872
Annual Costs ($M)			
Depreciation & Misc.	1,333	1,540	207
Operating & Maint.	570	711	141
Fuel	6,612	7,543	931
TOTAL	8,515	9,794	1,279
Power Cost (Mills/Kwh)	-	-	17.18
Before tax Return on Invest. (%)	-	-	27.1

Since the power is for local use, it is valued at system average industrial rate of 31.26 mills per kWh in determining the return on investment.

Figure 1.9 shows the pattern of return on investment for various levels of power generation, leading to the conclusion that it would probably be economically viable to use this method of filling all power requirements locally in those cases in which by-product power generation is viable by itself.

Potential Savings

In assessing potential savings, the point of view adopted assumes that whatever appears economically viable will be implemented, thus defining the full potential in the absence of other constraints. It is un-

FIGURE 1.9: ROI FOR VARIOUS LEVELS OF POWER GENERATION

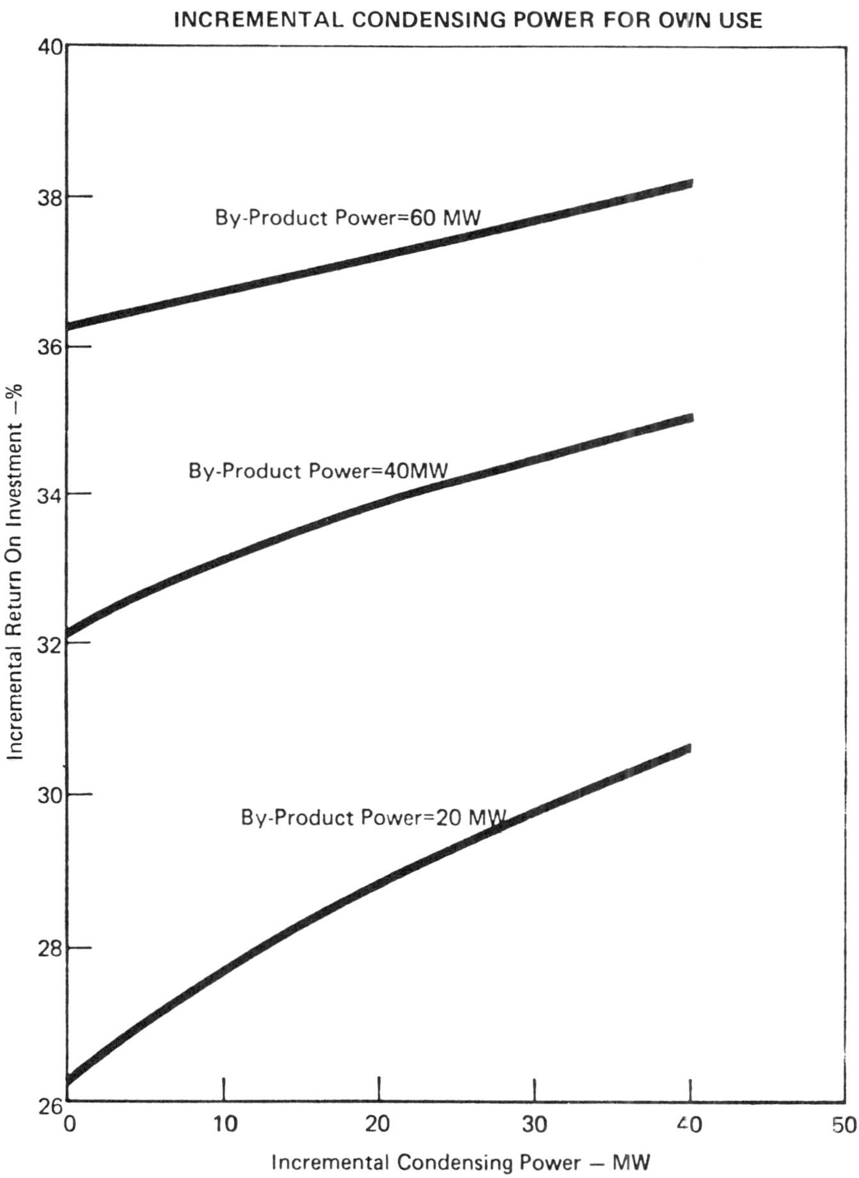

likely, of course, that this full potential would be realized as projected, if past milieu of political, emotional, traditional and other factors continues.

An evaluation of the 1980 projected costs indicates that industrial investment in by-product power facilities for own use would yield a 20% per year or greater before-tax return on the investment in excess of the return for steam alone, when the process steam load is 400,000 pounds per hour or greater. This steam rate was therefore taken as the lower limit for economic viability of by-product power generation. Adding incremental condensing power for own use to this by-product power base would yield before-tax incremental returns in excess of 20% per year. Adding still more incremental condensing power for sale would yield incremental before-tax returns of 20% per year or greater.

43% of the industrial steam load is generated in amounts of 400,000 pounds per hour or greater at single locations; if all of this were utilized for by-product power production, some 33,300 MW would be available by 1985. An additional 50% of incremental condensing power to bring the generation level up to full industrial electrical requirements would bring the generating capacity to 50,000 MW. If industry would be willing to invest an additional 20% of capital (over that needed to satisfy its own requirements) in more incremental condensing capacity, another 21,100 MW would be available for sale to utilities. The effect of this would be to relieve the utilities of $38 billion of investment, of which $19 billion would be a net national reduction in total investment. Net fuel savings in 1985 would amount to 1.45 quadrillion Btus, roughly equivalent to 680,000 barrels per day of oil. The fuel conversions involved would require an additional 320 million tons per year of 12,000 Btu per pound coal, of which roughly two-thirds is in addition to what the utilities would otherwise have consumed. If lower Btu content Western coal were used, the amounts would be proportionately higher.

[Editor's Note: The Dow Chemical Company has considerable experience in the field of cogeneration and meets a large proportion of its steam and electricity requirements through cogeneration.]

REFERENCES

(1) "Patterns of Energy Consumption in the United States," Stanford Research Institute, January 1972.

(2) William D. Witter Investment Strategy Weekly Report, November 26, 1973.

(3) 1967 Census of Manufacturers, U.S. Dept. of Commerce, Volume 1, *Summary and Subject Statistics.*

(4) *Lockwood's Directory of the Paper and Allied Trades*, Lockwood Publishing Co., New York City, 1974.

(5) *International Petroleum Encyclopedia,* Petroleum Publishing Company, Tulsa, Oklahoma, 1973.

(6) "Puerto Rico Energy Center Study," Burns and Roe, Inc. and The Dow Chemical Company, July 1970.

(7) "Industrial Complex Study for the Wasatch Front Area of the State of Utah," The Dow Chemical Company, May 1973.

(8) *Potential Fuel Effectiveness in Industry,* E.P. Gyftopoulos, L.J. Lazaridis, T.F. Widmer, Ballinger Pub. Co., Cambridge, Mass. 1974.

(9) "Fuel Conservation in Industrial Boilers," testimony by J.C. Wilcox, Jr. before the Senate Commerce Committee, June 10, 1974.

(10) "By-Product Power Rate," L. Elmenius, presented at the Fifth Annual Industrial Power Symposium, New Orleans, La., September 29, 1972.

(11) "Plant Sites 1975," John Winton, *Chemical Week*, p 33, October 23, 1974.

(12) "Steam-Electric Plant Construction Cost and Annual Production Expenses 1972," Federal Power Commission, April 1974.

(13) "Fossil Boilers as of 12/31/73," Clark, Dodge & Co., March 27, 1974.

(14) "Nuclear Reactors as of 12/31/73," Clark, Dodge & Co., March 1, 1974.

(15) Private communication from a major equipment manufacturer, June 27, 1974.

(16) "Energy Users Report No. 63," The Bureau of National Affairs, October 24, 1974.

(17) *ORCOST—A Computer Code for Summary Cost Estimates of Steam-Electric Power Plants,* ORNL-TM-3743, Oak Ridge National Laboratory, September 1972. The program used contained the cost model of October 1973.

(18) Federal Power Commission News Release No. 20523, July 29, 1974.

(19) *Generation Planbook 1974,* published by *Power* magazine, p 131.

(20) "Statistics of Privately Owned Electric Utilities in the United States—1972" Federal Power Commission S-236, December 1973.

(21) "Electric Utility Generating Equipment: Quarterly Status Report on Fossil Boilers," Kidder, Peabody & Co., November 5, 1974.
(22) "Economics of Higher Pressures and Temperatures for Steam Turbines in Industrial Plants," M.M. Patterson, E.V. Pollard, W.B. Wilson, presented at the 19th Annual Meeting of the American Power Conference, Chicago, Ill., March 27–29, 1957.
(23) "Theoretical Steam Rate Tables," J.H. Keenan and F.G. Keyes, American Society of Mechanical Engineers, New York, 1938.
(24) "Relative Engine Efficiences Realizable from Large Modern Steam-Turbine-Generator Units," G.B. Warren and P.H. Knowlton, ASME Semi-Annual Meeting, Milwaukee, Wis., June 17–20, 1940.
(25) *Steam Tables,* J.H. Keenan, F.G. Keyes, P.G. Hill, J.G. Moore, Wiley and Sons, New York 1969.
(26) "Linden Generating Station—Built to Help New Jersey Live Better ... Electrically," booklet by Public Service Electric and Gas Company, Newark, N.J. 1959.
(27) "The Midland Nuclear Process Steam/Electric Project," W.E. Kessler and J.R. Burroughs, presented at the First National Topical Meeting on Nuclear Process Heat, Los Alamos, N.M., October 1–3, 1974.

Thermo Electron Corporation Study

The material in this chapter is based upon PB-255 659, entitled *A Study of Inplant Electric Power Generation in the Chemical, Petroleum Refining and Paper and Pulp Industries,* prepared by Thermo Electron Corporation for the Federal Energy Administration, Office of Energy Conservation, published in June 1976. The authors are S.E. Nydick, J.P. Davis, S. Fain, and R. Sukhuja. The sections of PB-255 659 used in this chapter concern the costs of power systems and the economics of on-site power generation. Sections of this report on industrial data bases, technology, and potentials for fuel savings in the industries studied have been omitted, since these subjects are covered in Chapters 3 and 4 of this book with more recent data. A summary of the conclusions reached by Thermo Electron Corporation, however, has been included, and is quoted verbatim.

INTRODUCTION

Numerous studies have documented the end uses of fuel in U.S. industries. A most important conclusion drawn from these studies is that about 45%, 13×10^{15} Btu/yr, of all the fuel consumed in U.S. industries is used for steam generation, chiefly at low pressure. An additional 29%, 8.3×10^{15} Btu/yr, is consumed in fossil fueled direct heating processes. The remaining 26% is consumed for the generation of electricity for electrical heating, motors, lighting, electrolysis, etc. Only about 15% of the electricity consumed by industry and 5.8% of that consumed nationwide is generated inplant with the remainder generated by utilities at a nationwide average heat rate of 10,500 Btu/kWh with condensing steam turbines or gas turbines without heat recovery. This heat rate can be compared to heat rates of 4,000 to 7,000 Btu/kWh obtainable from steam turbine, gas turbine or diesel topping of industrial process steam and utilizing the same basic technology used by utilities. The nationwide potential for inplant generation of electricity varies with the topping cycle. For example, at approximately 50 kWh of electric power per million Btu of process steam, steam turbine topping can produce about 35% of all the electricity generated nationwide at a heat rate averaging 4,500 Btu/kWh. Gas turbine topping can generate nearly 200 kWh and diesel topping 400 kWh per million Btu of process steam or more electrical power than is produced in the U.S. as a heat rate in the vicinity of 5,500 and 7,000 Btu/kWh, respectively.

Better than 30% of energy input into fossil fuel direct fired heaters eventually leaves through the stack gases. With bottoming cycles recovering this waste heat, about 6.7% of the U.S. electricity can be potentially generated fuel free. Accumulatively, steam and gas turbine topping and steam or organic bottoming cycles have the potential of saving the U.S. the fuel equivalent of almost 5.0 million barrels of oil per day, which is 15% of all the fuels and 32% of all the petroleum consumed.

Because of economic, regulatory, and institutional barriers, this potential for fuel savings has not been tapped in the U.S. In fact since 1950, when 15% of the nation's electrical power was generated inplant, the percentage thus generated has steadily declined to nearly 5% in 1973. The post war period of cheap fuel, technological advances in more efficient and large utility power plants, and the design of economical packaged boilers for industrial steam process use were among the most important economic factors contributing to the decline of inplant generation. Industrial rate structures with high demand and emergency charges, other regulations which prevented or impeded inplant generation and wheeling of power, tax regulations, and the opposing interests of utilities and industry also contributed to this demise.

Because of the impending shortage of fuels, it is within the national interest to investigate and determine if some of these barriers have disappeared and how they can be overcome, so that the potential for nationwide fuel savings by means of inplant generation can be achieved. Surely, as a result of the more than threefold increase in fuel prices since 1973, the economics of inplant versus utility generation have been altered. Extrapolating from industrial experience in Germany where fuel prices were high before 1973 and inplant generation amounts to about 29% of total electrical generation, economic justification should now exist for the realization of a substantial amount of the potential for inplant generation. With favorable economics, the legal, regulatory and institutional barriers to inplant generation that existed previously should be substantially easier to overcome since it will be in everyone's self interest—the utility, the industrial plant, the consumers, and the federal government—to exploit this energy conservation strategy.

POWER SYSTEM COSTS

The 1975 installed system and operation and maintenance costs of power systems for inplant generation of electricity are discussed in this section. The installed costs for topping systems for the combined generation of steam and electricity were calculated for three situations:

1) New installations and expansions of plant and replacement of old and worn-out equipment.
2) Replacement of existing and useful, low-pressure boilers with remaining service life.
3) Installation of a back pressure steam turbine between the point of generation and the point of use where the pressure between these two points is currently reduced in a valve.

For new installations, etc., there is a choice between a topping cycle producing steam and electricity in combination (by industry or utility) or separate generating cycles for steam (by industry) and electricity (by utilities). An incremental analysis (i.e., the cost attributable to generating the electricity in the topping cycle) showing the differences between these two alternatives is presented and will be utilized for the evaluation of the economics of new installations. For the replacement of existing and useful equipment, the total installed costs for the topping cycle equipment is utilized in economic analysis since the plant could continue to generate its process steam with its present low-pressure boiler without any capital outlay. For similar reasons the total installed cost is also used in the economic analysis of back pressure turbine installations. Equipment costs for 1975 were obtained mainly from a study performed by Stone and Webster for this program. Calculations were performed for power systems as a function of plant size for the purpose of defining economically viable plant sizes applicable to various topping and bottoming power systems.

Steam Turbine Topping

A schematic of an elementary steam turbine cycle is shown in Figure 2.1. The primary system components for steam turbine topping are the boiler, steam turbine, pollution control, and fuel handling equipment. The installed costs for these system components are shown in Figures 2.2 , 2.3 and 2.4. Boiler installed costs include the boiler and erection, foundations, building, piping, instruments, auxiliary equipment, ductwork and stack, electrical material in boiler building, feedwater sampling and treatment, fuel and ash handling and indirect and distributable costs. The ratio of installed to boiler costs for oil-fired and coal-fired boilers ranged from 2 to 2.5 and 1.9 to 2.1, respectively, depending on boiler size.

For steam turbines the direct costs were multiplied by 2.25 to obtain installed costs, which include erection, piping, electrical accessories, foundations, building, instruments, insulation, indirect and distributable costs. Costs for coal handling and storage and ash removal, and oil handling and storage equipment were obtained by applying a 0.6 power law to estimates made by L.M. Olmsted in *Electrical World* of November 11, 1973, page 39, for a 400 MW power plant ($10.73/kW for coal handling and ash removal and $6.22/kW for oil handling). Electrostatic precipitators were included as the strategy for particulate control in both coal and oil-fired boilers. Electrostatic precipitator costs were based on an installed cost of $5.00/acfm at 140,000 acfm, 550°F, and a 0.75 power law for other plant sizes. Sulfur removal costs were estimated at about $1.00/lb/hr of steam (equivalent to about $100/kW for a condensing steam plant).

Initial calculations for total and incremental installed costs were performed as a function of turbine inlet pressure as well as turbine discharge pressure and plant size. However, it quickly became apparent that, for most applications, 1,250 psig would be the optimum (or very close to the optimum) turbine inlet pressure economically (i.e., yielding the greatest internal rate of return) and would result in maximum electrical production. Calculations for total and incremental installed costs were thus limited to 1,250 psig/900°F turbine inlet. Because there

FIGURE 2.1: STEAM TURBINE TOPPING FOR COMBINED GENERATION OF ELECTRICITY AND PROCESS STEAM

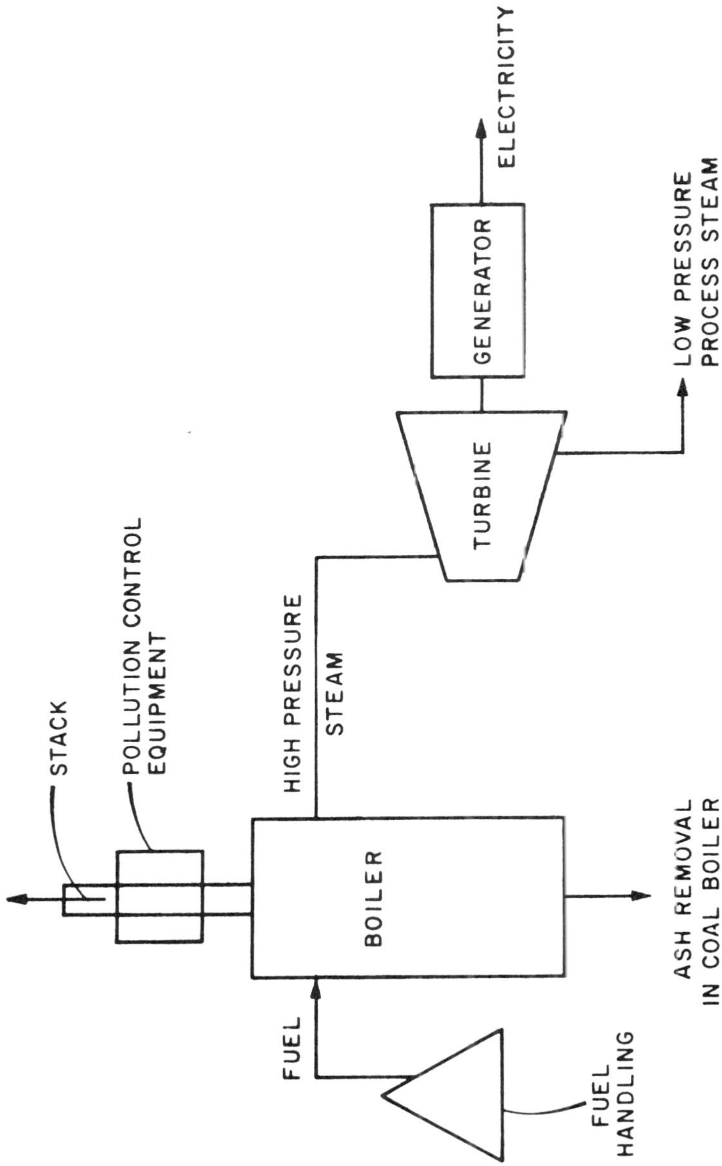

FIGURE 2.2: INSTALLED COSTS (1976 DOLLARS) vs CAPACITY
FIELD ERECTED STEAM BOILERS

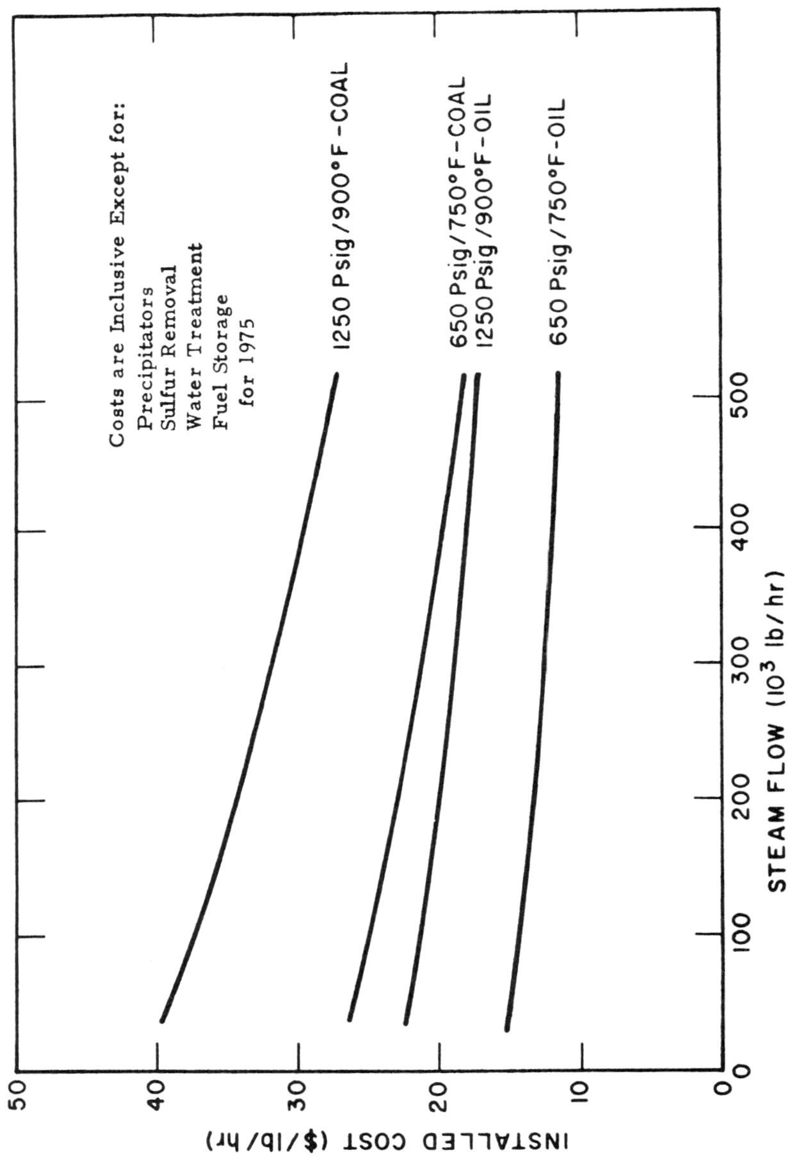

FIGURE 2.3: INSTALLED COSTS (1976 DOLLARS) vs CAPACITY BACK PRESSURE TURBINE GENERATORS

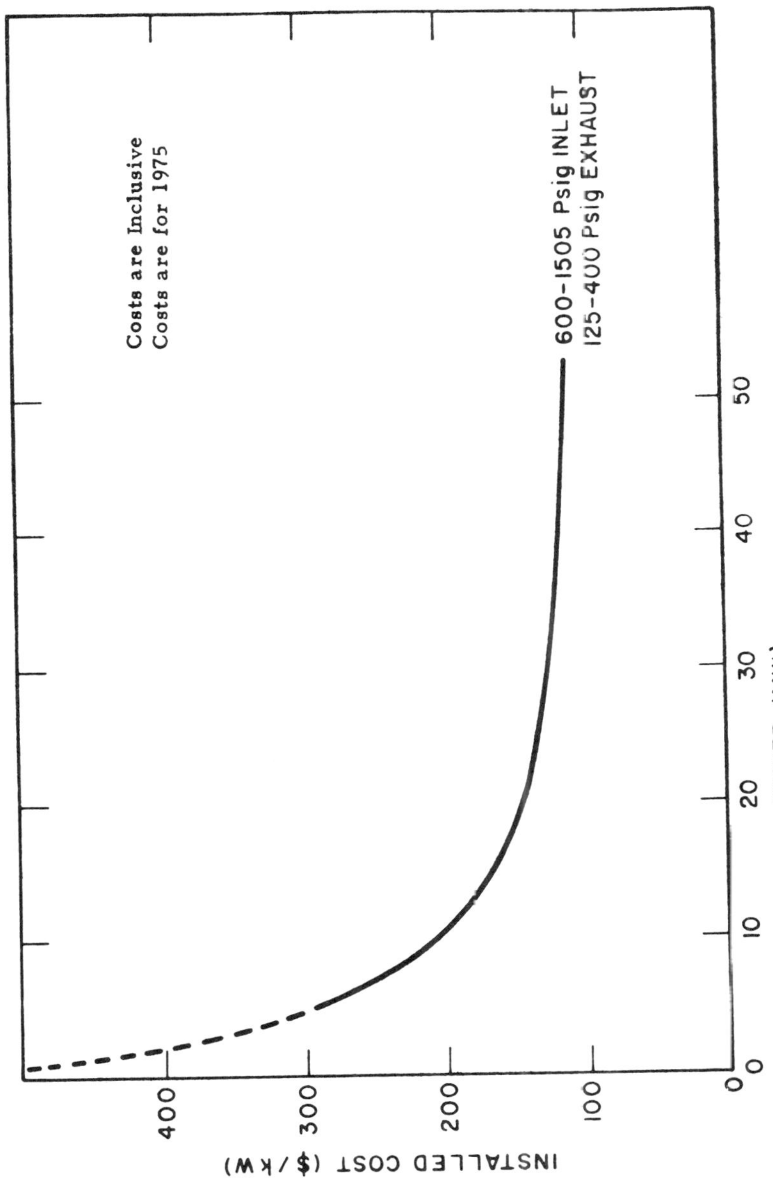

FIGURE 2.4: INSTALLED COSTS FOR COAL AND OIL HANDLING AND EMISSION CONTROL STRATEGIES

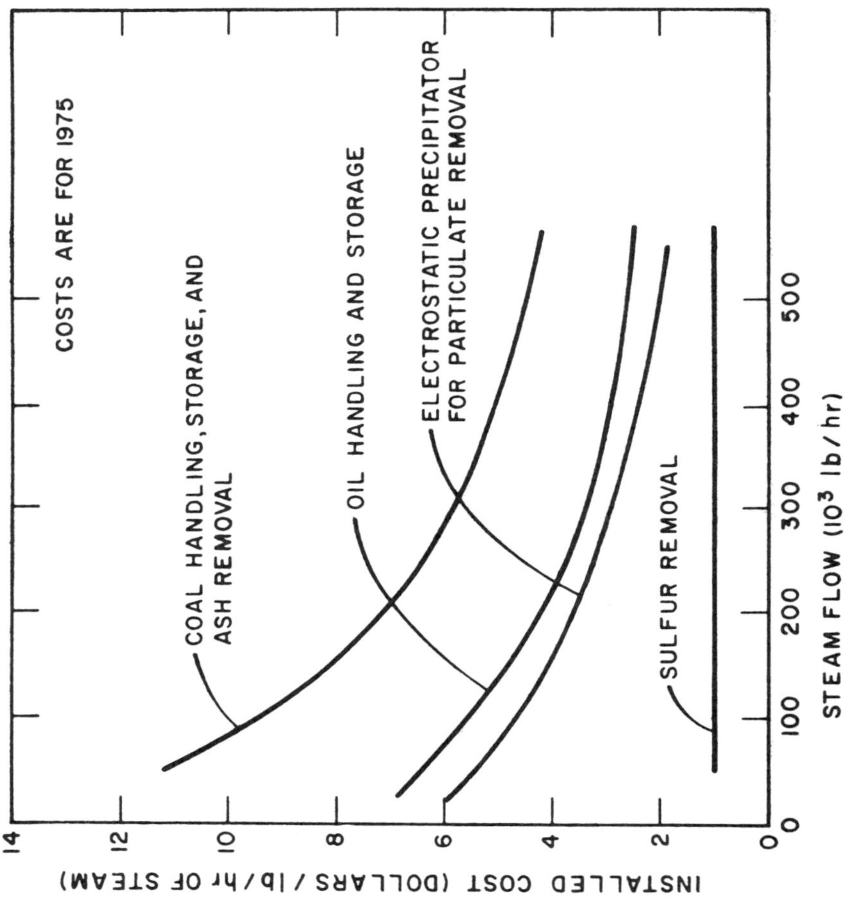

are many existing boiler installations generating steam at pressures as high as 650 psig and reducing the steam pressure for process conditions without any production of useful work (electricity or shaft horsepower), installation of a back pressure steam turbine between a 650 psig/700°F steam source and process steam conditions was also considered.

Total installed costs for steam turbine topping for 350, 150, and 80 psig process steam conditions are shown in Figure 2.5 vs plant size. The total installed costs in Figure 2.5 include the installed cost of the boiler, steam turbine, fuel and ash handling, and pollution control equipment. Incremental installed costs, representing the cost of the steam topping system less process steam boiler are shown in Figure 2.6. Incremental costs exclude pollution control and fuel and ash handling costs since it is assumed that both high-pressure (for the topping cycle) and low-pressure (the other alternative choice) process steam boilers require these items. Installed costs for a back pressure turbine located between an existing 650 psig boiler and process steam use is shown in Figure 2.7. Operation and maintenance (O & M) costs for steam turbine topping systems were assumed to be $0.003/kW.

Gas Turbine Topping

Installed costs for the gas turbine generators were obtained from the Stone and Webster study and from available industry data; they include an installed cost factor of 1.6 to account for installation labor, accessory equipment and piping, foundations, turbine building, instruments, insulation, painting, indirect costs, distributable costs. The factor does not include transformers, escalation interest during construction, land, or other owner's costs. Waste heat boiler costs (Figure 2.8) from the Stone and Webster report and fuel storage installed costs (Figure 2.4) were added to the generator costs to obtain total installed costs (Figure 2.9). In general, the installed costs of heat recovery boiler systems depend on steam pressure and flow rate and include an installed cost factor of 2.5 to cover the cost of foundations, ductwork, and piping stack, instruments, insulation, painting, indirect costs and distributable costs. As before, no escalation interest during construction, land, and other owner's costs have been included. Characteristics of the Rolls Royce model RB211-24 gas turbine plant were used as a guide in computing turbine power as a function of processing steam rate.

In contrast to the steam turbine topping, the incremental cost (Figure 2.10) was obtained from the total installed cost by subtracting the boiler

FIGURE 2.5: STEAM TURBINE TOPPING TOTAL INSTALLED COSTS* vs STEAM FLOW

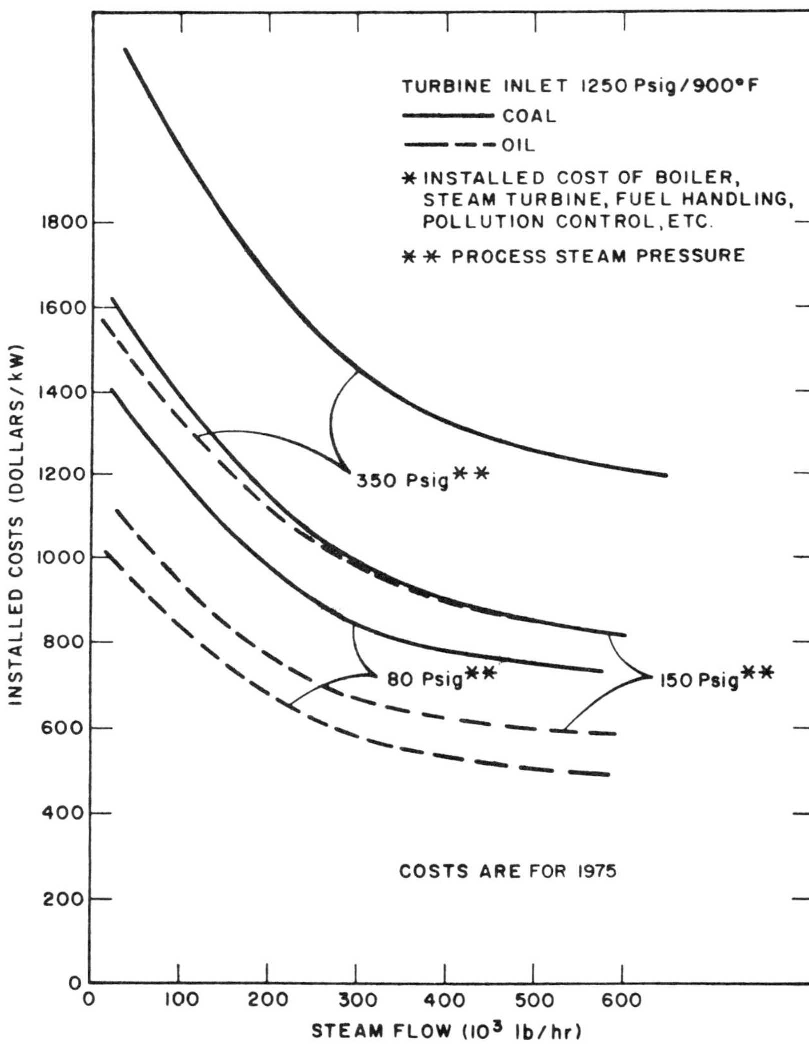

FIGURE 2.6: STEAM TURBINE TOPPING INCREMENTAL INSTALLED COSTS* vs STEAM FLOW

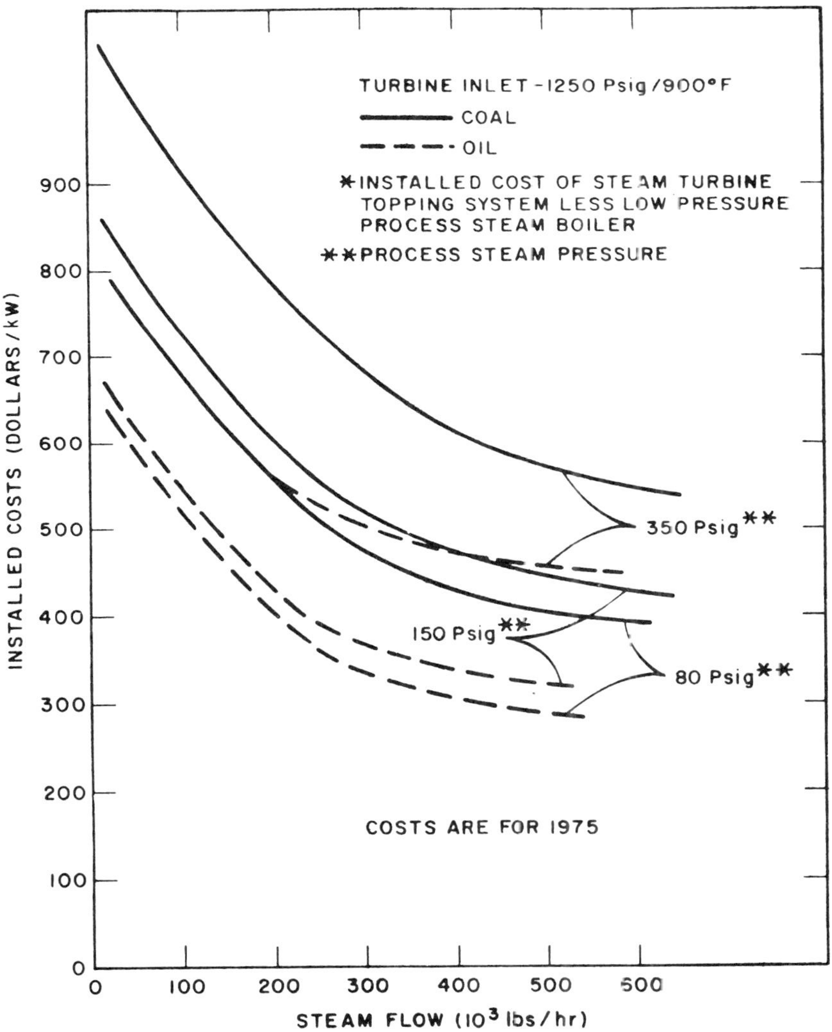

FIGURE 2.7: INSTALLED COSTS* FOR INSTALLATION OF BACK PRESSURE TURBINE vs STEAM FLOW

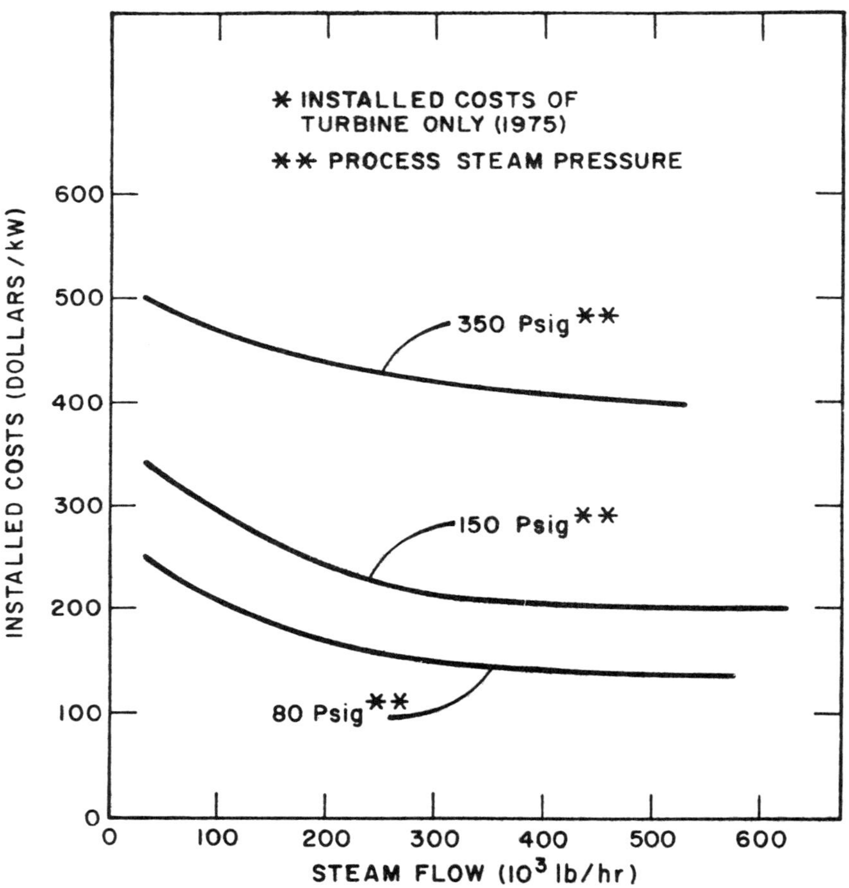

FIGURE 2.8: INSTALLED COST FOR HEAT RECOVERY BOILERS

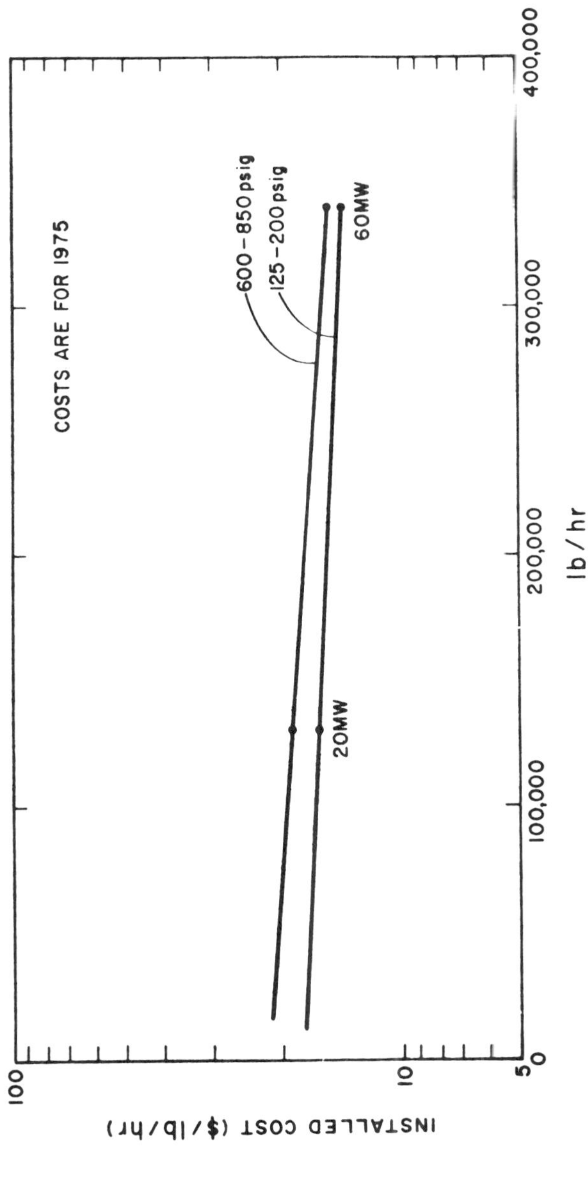

FIGURE 2.9: GAS TURBINE TOPPING INSTALLED COST vs STEAM FLOW

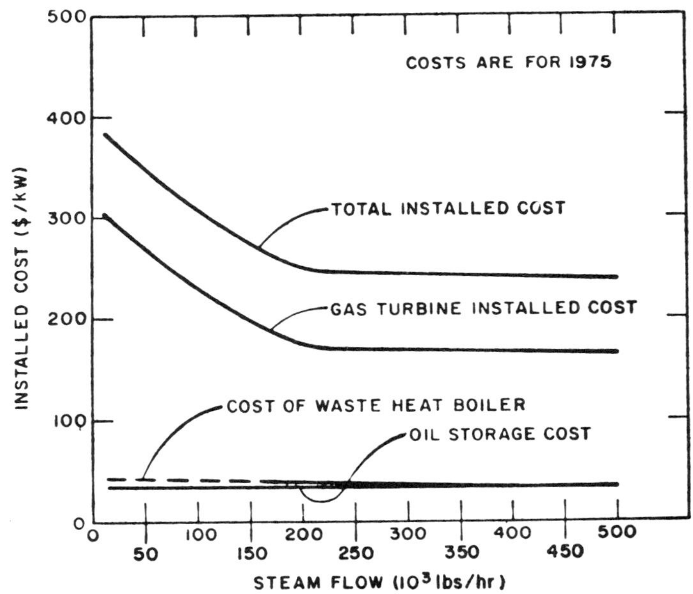

FIGURE 2.10: INCREMENTAL COST vs STEAM FLOW

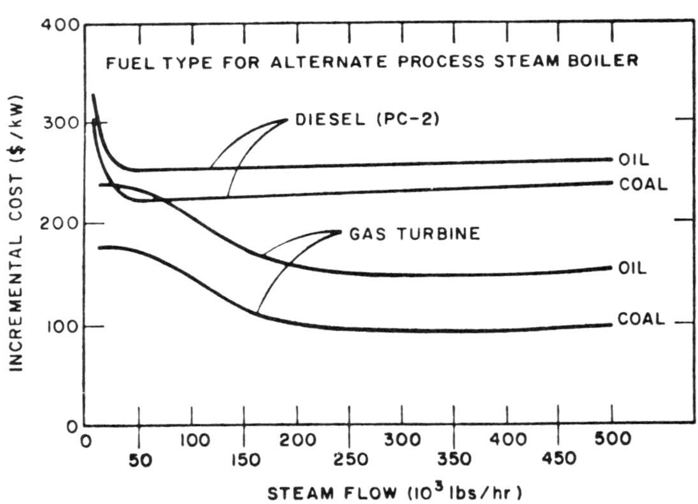

FIGURE 2.11: DIESEL TOPPING TOTAL INSTALLED COST vs STEAM FLOW

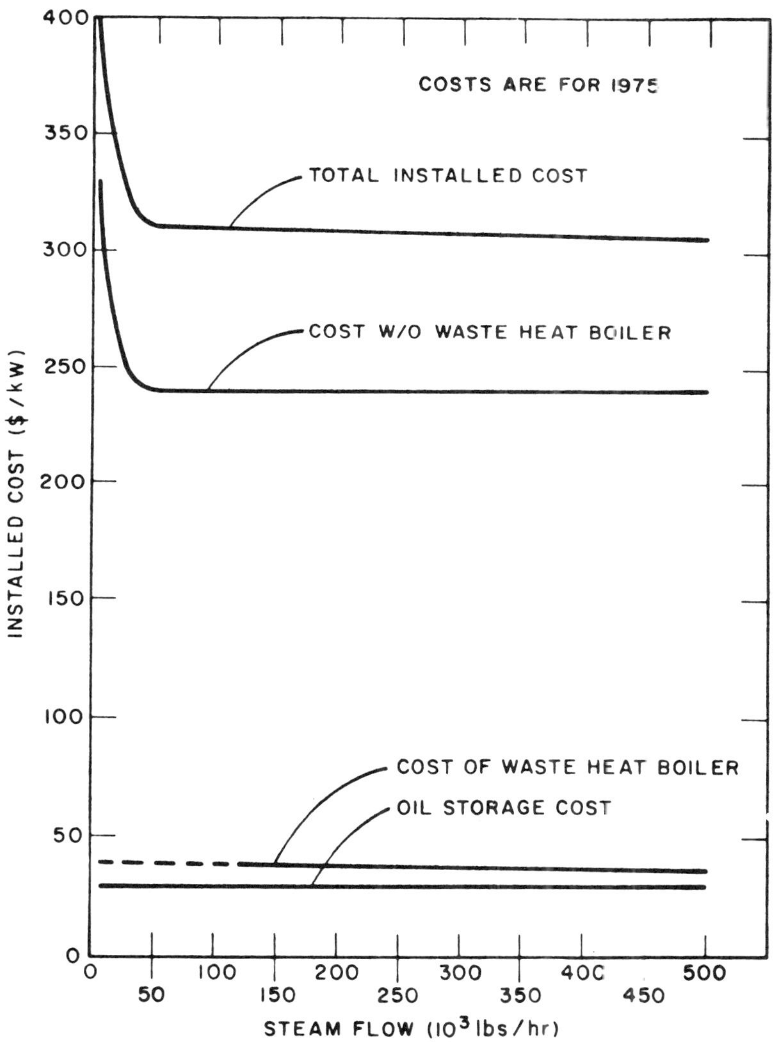

FIGURE 2.12: ORGANIC RANKINE AND STEAM BOTTOMING TOTAL INSTALLED COST vs WASTE HEAT AT TEMPERATURE

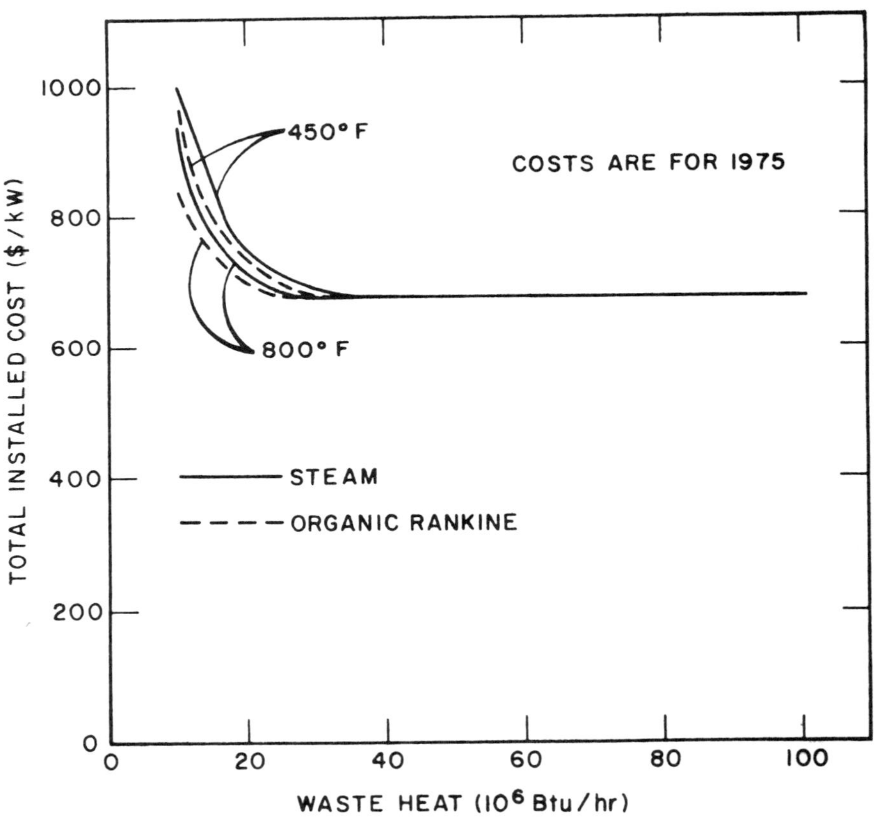

and attendant pollution control and fuel and ash handling costs since, in this case, no process steam boilers are actually in existence.

Also, in contrast to the steam turbine topping, the cost effect of process steam pressure in the 100 to 400 psig range is negligible owing to the much higher electric power generated relative to process steam flow. O & M costs of $0.004/kWh were assumed for a plant duty cycle of 0.8.

Diesel Topping

Characteristics of the Fairbanks Morse PC2 Diesel were used in computing topping power as a function of process steam rate. Installed costs for the diesel generators were obtained from industrial data with the same installed cost factor as for the gas turbines (Figure 2.11). Incremental costs (Figure 2.10) subtract the cost of an alternative process steam boiler and associated pollution control and fuel and ash handling costs. As in the case of gas turbine topping the cost effect of steam pressure in the 100 to 400 psig range is negligible because of the high ratio of electric power generated to process steam flow. O & M costs of $0.0033/kWh were assumed for a plant duty cycle of 0.8.

Low-Pressure Steam and Organic Rankine Bottoming

An attractive feature of bottoming plants is the generation of electric power without fuel consumption. Hence, the installed cost of the bottoming plants includes only the installed cost of the waste heat boiler and the turbine generator. A pinch point of 40°F was assumed to compute process steam conditions at the exit of the counterflow waste heat boiler. Figure 2.12 illustrates the total installed costs of the steam and organic Rankine bottoming plants as a function of waste heat for two representative temperatures. The lower cost of the organic Rankine plant results from the fact that it converts a higher fraction of the available waste heat into electrical output with the same plant cost characteristic. O & M costs of $0.003/kWh were assumed for both plants at a duty cycle of 0.8.

ECONOMICS OF INPLANT POWER GENERATION

Investment Decision Making

Internal Rate of Return: The basic process of capital spending analysis consists of evaluating the investment outlay in terms of the economic gain provided by it. The investment outlay is represented by the net investment required for the new equipment or change in operation, and the economic gain is represented by estimated operating cash flows generated by the investment required for inplant generation. Since capital spending occurs early in an improvement program and the economic gain is realized in later years, a common time must be used for comparison of these funds.

For these comparisons, the internal rate of return (or discounted cash flow) method of analysis will be used, which is based on the criterion that the sum of the present value of all cash flow returns associated with a given project is equal to the initial company expenditures, namely:

$$C_o = \sum_{n=1}^{N} \frac{\text{Net cash flow (n)}}{(1 + i)^n} \tag{1}$$

where: C_o = capital expenditure by investing company
n = rate of return
N = economic life in years

The factor $1/(1 + i)^t$ transforms each expenditure or cash flow return to its value at time zero. The net cash flow in year, t, is defined as the savings (or, for utilities, the revenues resulting from the sale of electricity) resulting from the expenditure, i.e., savings resulting from a reduction in purchased electricity, minus the cost of the fuel chargeable to the production of the electricity, operation and maintenance (O & M) costs, federal taxes, and loan payments.

$$NCF_n = S_n - FC_n - OM_n - FT_n - ALP_n \tag{2}$$

where: NCF_n = net cash flow for year n
S_n = savings on sales of electricity for year n
FC_n = fuel costs for year n
OM_n = operation and maintenance costs for year n
FT_n = federal taxes for year n
ALP_n = annual loan payments for year n

The federal taxes are calculated by the following:

$$FT_n = TR \times (S_n - FC_n - OM_n - D_n - LI_n) \tag{3}$$

where: TR = tax rate

D_n = depreciation for year n

LI_n = interest on loan for year n

It can be assumed that the gross savings, GS, for year n, defined as $(S_n - FC_n - OM_n)$ escalates at a rate, r, per year. Thus, the gross savings can be expressed as:

$$GS_n = (S_o - FC_o - OM_o)(1 + r)^n \quad (4a)$$

$$= GS_o (1 + r)^n \quad (4b)$$

where: GS_o = gross savings yearly at time of investment.

Combining equations 2-4, we obtain:

Year n

$$NCF_n = (1 - TR)(1 + r)^n GS_o - TR(D_n + LI_n) - ALP_n \quad (5)$$

If the investment is made at year end, the investment tax credit can be subtracted from the total investment and equation (1) becomes:

$$(1 - I)\, C_o = \sum_{n=1}^{N} \frac{1}{(1 + r)^n} \left[(1 - TR)(1 + r)^n N_o S_o - TR(D_n - LI_n) - ALP_n\right]$$

Gross Payback Period: Gross payback period, defined as the ratio of capital expenditures for project to yearly pre-tax gross savings at time of investment (C_o/GS_o), is often used to evaluate investment projects. The validity of gross payback period as a criterion for investment is based upon its relationship to the internal rate of return. For a given investment opportunity, gross payback is the independent variable, and internal rate of return the dependent variable. Changes in tax regulations, depreciation, accounting, financing procedures, escalation in savings and costs, etc., will also change the internal rate of return; however, the gross payback is independent of these variables. Thus, in discussing different tax regulations, etc., gross payback is useful only as a computational tool. Internal rate of return becomes the meaningful investment criterion.

Required Rates of Return on Investment: The required rate of return on investment (ROI) varies with the company, industry, and general economic climate. An average ROI, based upon the weighted averages of debt and equity, value line risk factors, bond ratings, and riskless borrowing rates, can be obtained for an industry for all its investments. The industry-wide average after tax rates of return calculated from these parameters are 15, 16, 15 and 13% for the chemical, petroleum (the integrated industry), and paper and pulp industries and for utilities, respectively, according to Mathematica Inc.

Historically, inplant generation of electricity has been considered by industry, except in isolated cases, as an ancillary investment requiring higher rates of return, since it is not associated with the profitability or expansion of its primary market. The integrated paper and pulp mills, which nationally produce 55 to 60% of its electrical requirements inplant, and one or two of the largest chemical and petroleum refineries are the most notable exceptions. The large amount of inplant generated power in paper and pulp mills is largely the result of the availability of substantial quantities of waste bark and other wood products and spent liquor (contributing 40 to 50% of its total boiler fuels) and the need for large amounts of steam (about 90% of their total energy requirements).

Thus, for industry a substantial fraction of companies, perhaps the majority, will not invest in inplant generation projects with internal rates of return of about 15%, generally judged to be a satisfactory return for product oriented investments. The accumulative distribution of companies which will invest is assumed to be represented by Figure 2.13. From this figure about one-fifth of the companies in the chemical, petroleum refining, and paper and pulp industries will invest in inplant generation at a return of 15%, whereas four-fifths would invest at a return of 40% (a gross payback of 2 years). Generation of electricity, however, is the utilities' primary product line and therefore utilities will invest at rates of return near the calculated average return, regulated by state utility commissions since they are regional monopolies.

In addition, utility borrowing rates have traditionally been lower than for industry. Thus, from Figure 2.13 it is estimated that 50% of the utilities would invest at an internal rate of return of 13%, the average calculated required return, and 90% at a return of 18%. By comparing the internal rate of return of inplant generation to utility investment criterion, we are actually comparing the profits derived from inplant generation with utility plant generation, since an average internal rate of return of 13% for utilities implies that this is the average return derived from central station condensing power. The investment criterion for utilities in Figure 2.13 implies, therefore, a distribution of the rate of return earned by central station condensing power plants.

In the analysis that follows, for simplicity, the same cash flows and capital costs (i.e., same gross payback) were used for industry and util-

FIGURE 2.13: ACCUMULATIVE DISTRIBUTION OF INDUSTRY AND UTILITIES WILLING TO INVEST vs RATE OF RETURN

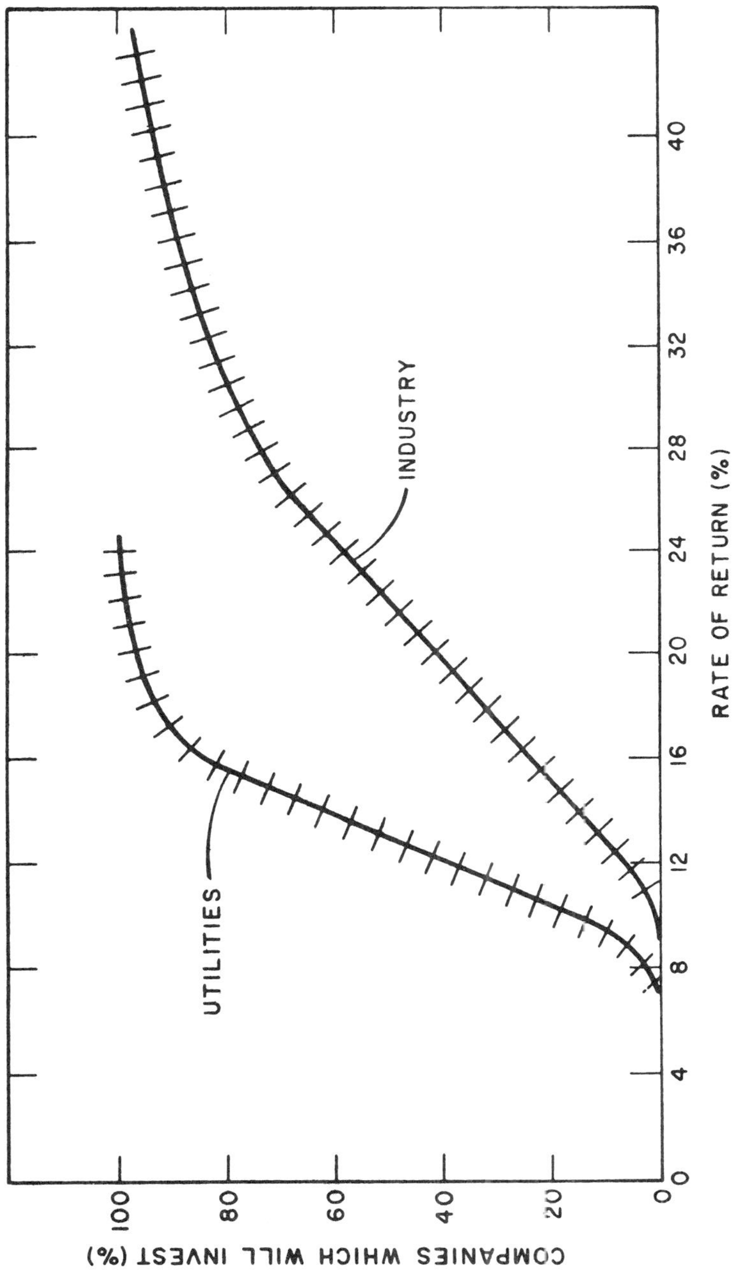

ity investment. This implies certain assumptions for utilities. For a topping system, cash flows for utilities would result from the sale of electricity and steam, while for industry they would result from the purchase of less electricity (both utility charge and industry cost per kWh of electricity assumed to be the same). In addition, the fuel cost for utilities would be based on the total fuel used for the topping plant (i.e., fuel chargeable to steam plus fuel chargeable to power) rather than the component chargeable to power that is used for industry. Using the same cash flows for both analyses thus assumes that the utility is selling the steam for the cost of the fuel chargeable to the steam (i.e., revenues from steam flow balance the costs of producing process steam).

Since it is most likely that utilities will make a profit on steam sales, this is a conservative assumption. Thus, in a replacement analysis (of usable equipment), using the total installed cost of a system for utilities implies that the industry is receiving process steam from the utility at a cost equal to its own cost of producing steam in its presently usable boilers and the utility is paying the capital cost increment for the process steam (which, when judged on a dollars/kW basis, may be greater than for a central plant station) with the subsequent advantage of reducing its fuel cost for the production of electricity.

On the other hand, the incremental analysis for utilities assumes that: (1) industry will contribute that portion of the capital cost attributable to process steam (i.e., low pressure steam boilers) to the total system cost with the utility contributing the remainder of the cost (i.e., that portion attributable to power), or (2) the utility is including the cost of capital attributable to steam production in its charge for the steam. Thus, for the incremental analysis, industry is willing to pay for the incremental capital and fuel cost of producing steam while the utility is basing its investment on the incremental (capital and fuel) cost of producing power. Both the total and incremental analysis applied in this study to utilities is conservative (i.e., resulting in a lower return of investment) since the utility will probably make money on the steam as well as on the electricity produced from a topping plant. Since bottoming plants produce one product, electricity, the difficulties encountered above in comparing analyses for industry and utilities do not occur.

Effect of Income Tax Incentives and Financing Techniques on Investment: Changes in income tax regulations and financing can significantly alter the internal rate of return of a prospective project to the investing company; i.e., it can significantly change the money a company can expect to make and thus the probability of companies investing. Figure 2.14 compares the effect of altering the investment tax credit and income tax rate on internal rate of return as a function of gross payback period (capital costs/pre-tax savings at time of investment). It may be seen from this figure that for an investment resulting in a payback of 10 years, the internal rate of return will increase from about 11% to 21% with an increase in the investment tax credit from the present 10% to 50% (50% tax rate). Similarly, the internal rate of return increases to about 14% for a 25% tax rate (10% investment tax credit). Figure 2.15 shows a similar comparison with different financing. Financing half the capital requirements (at 10% interest with constant yearly interest and principal payments over 20 years) increases the internal rate of return for a 10-year gross payback from 11% to 17%.

The effect of tax and financing charges on the probability for investment can be seen clearly by combining Figure 2.13, showing the accumulative fraction of companies which are likely to invest as a function of internal rate of return, and Figures 2.14 and 2.15 showing the effect on internal rate of return. Figures 2.16 and 2.17 combine these two relationships for industry and utility investment criterion, respectively. We see from Figure 2.16 that for a 10-year gross payback, 5%, 27% and 45% of industry will invest with present tax regulations and 0% financing, with present tax regulations and 50% financing, and 50% investment tax credit and 0% financing, respectively. Similarly, from Figure 2.17, we see that for a 10-year gross payback, 50%, 92% and 98% of utilities will invest with the same respective tax and financing conditions.

COST/BENEFIT OF INPLANT GENERATION

In this section, results of calculations for the internal rate of return, and the corresponding percent of industry and utilities that will invest (based upon the analysis described in Power Section Costs) for the different power systems and for installation between 1975 and 1985, will be presented for different system parameters. It is of particular significance to know the minimum plant size in which industry or utilities are willing to invest. Capital costs for 1975–1985 installation were obtained by averaging the expected 1985 costs (from construction cost index) with the calculated 1975 costs. Installed and incremental costs in 1975 for these power systems were discussed previously. Regional fuel and electrical costs for 1975 and 1985 are shown in Table 2.1. Average fuel costs and electrical costs for 1975 and 1985 were used in the calcula-

FIGURE 2.14: INTERNAL RATE OF RETURN vs GROSS PAYBACK—EFFECT OF INVESTMENT TAX CREDIT AND TAX RATE

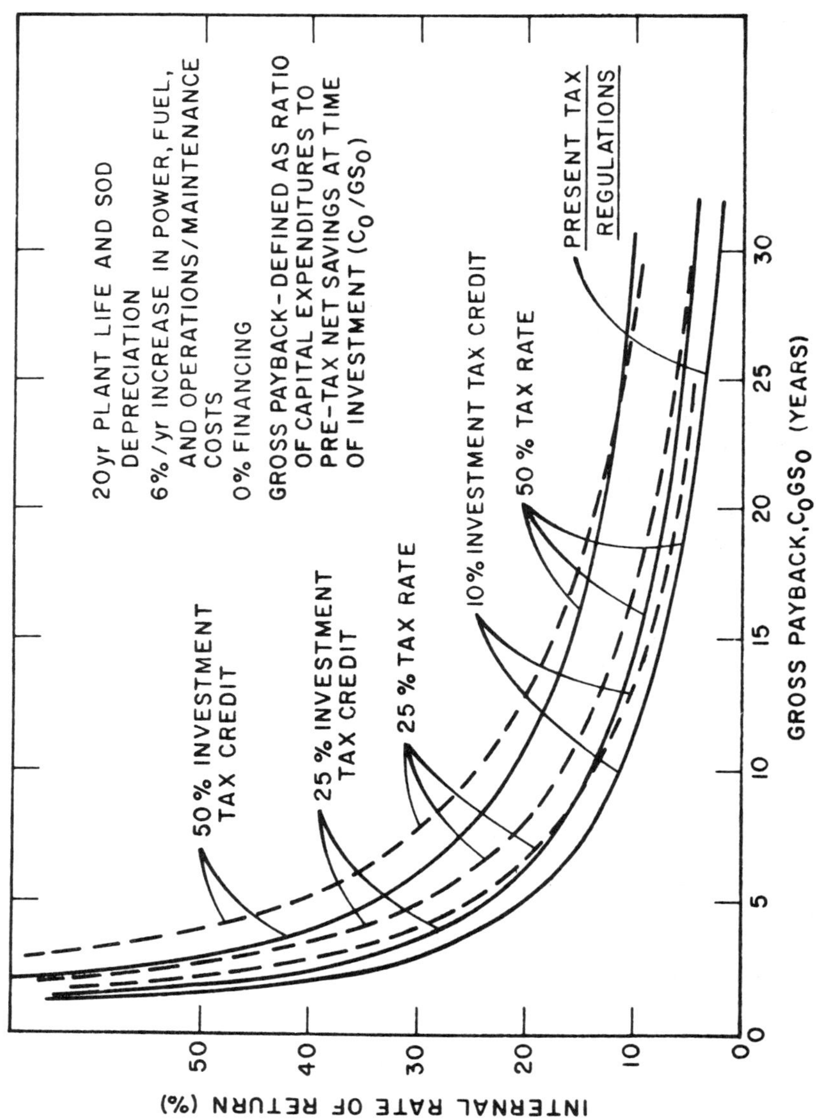

FIGURE 2.15: INTERNAL RATE OF RETURN vs GROSS PAYBACK—EFFECT OF INVESTMENT TAX CREDIT AND FINANCING

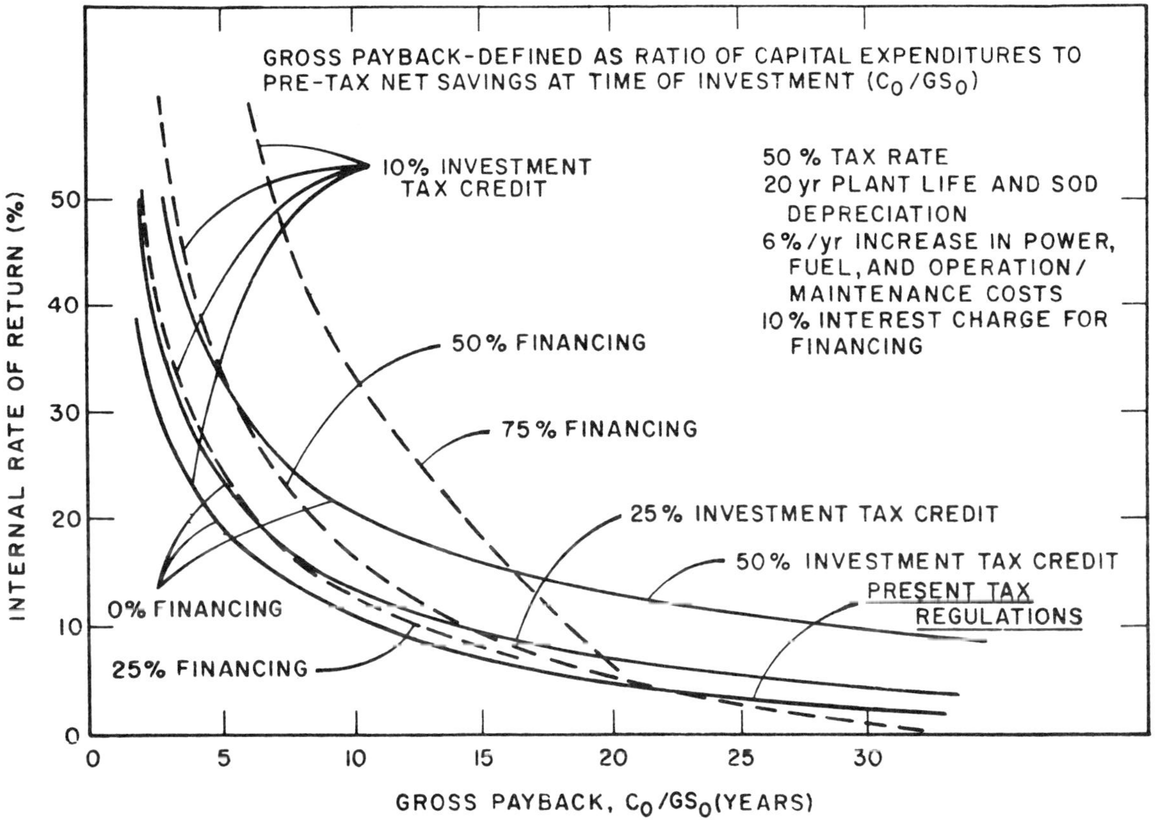

FIGURE 2.16: ACCUMULATIVE DISTRIBUTION OF INDUSTRY WILLING TO INVEST vs GROSS PAYBACK

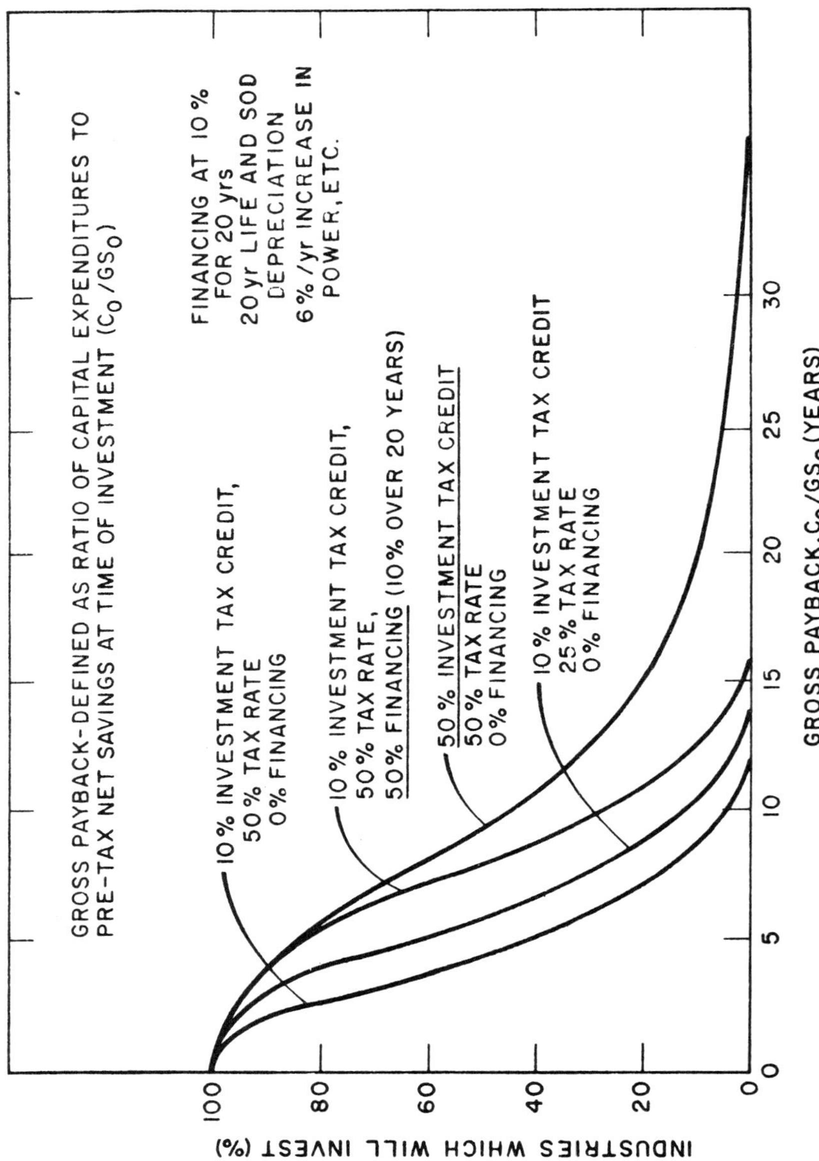

FIGURE 2.17: ACCUMULATIVE DISTRIBUTION OF UTILITIES WILLING TO INVEST vs GROSS PAYBACK

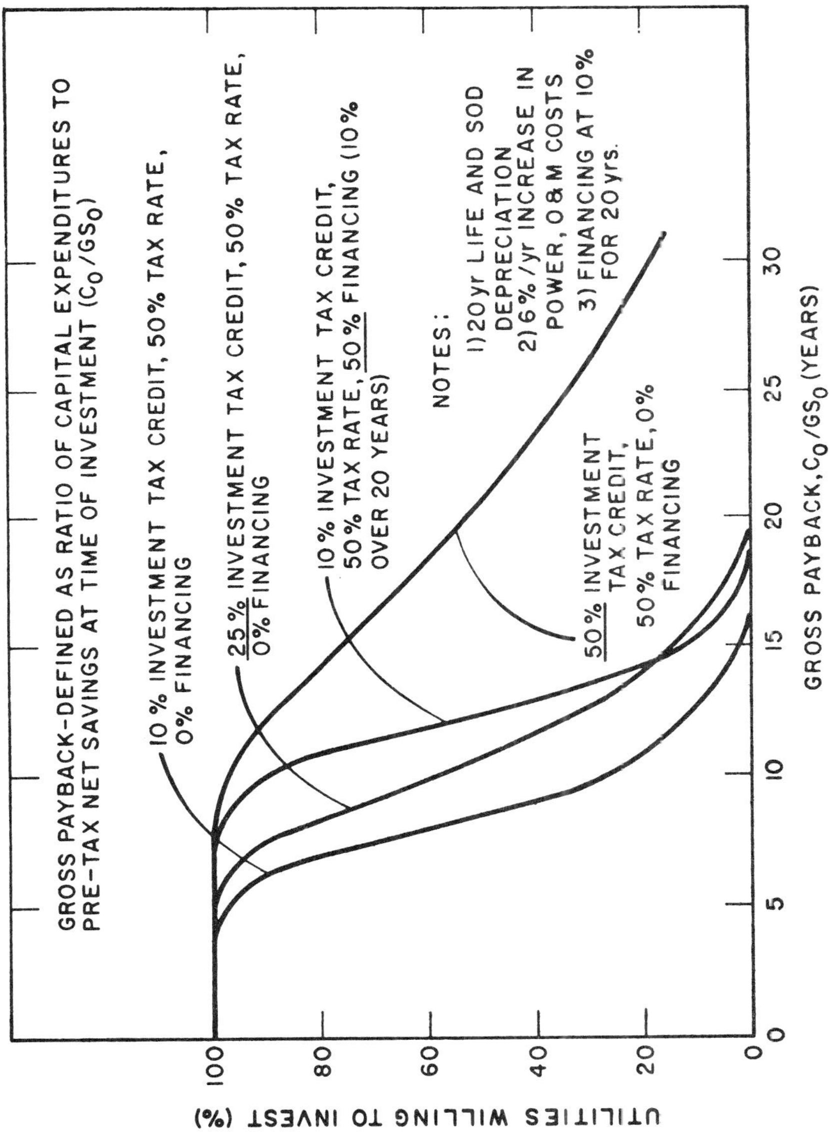

TABLE 2.1: REGIONAL FUEL AND ELECTRICITY COSTS

Region	Coal		Oil		Gas		Electricity	
	1975	1985	1975	1985	1975	1985	1975	1985
	Dollars/10^6 Btu		Dollars/10^6 Btu		Dollars/10^6 Btu		Cents/kWh	
New England	1.98	2.98	2.49	3.98	1.84	3.14	3.15	4.15
Middle Atlantic	0.90	1.35	2.31	3.70	1.11	2.10	3.14	5.5
East North Central	0.80	1.07	1.77	2.63	0.93	2.18	2.36	4.9
West North Central	0.60	0.78	1.43	2.21	0.71	2.71	2.14	4.4
South Atlantic	0.93	1.30	2.01	3.21	0.94	2.29	2.52	5.2
East South Central	0.72	0.96	1.25	2.07	0.70	1.93	2.08	5.7
West South Central	0.46	0.60	1.78	2.95	0.70	2.61	1.55	4.2
Mountain	0.45	0.67	3.02	5.02	0.64	2.04	1.64	4.7
Pacific	0.68	0.93	1.91	2.83	0.89	2.46	1.78	5.5

Source: Coal, oil and gas costs - Mathematica Data Base

Electricity - Thermo Electron and Mathematica Data Base

tion of the savings resulting from inplant generation for installation between 1975 and 1985.

Steam Turbine Topping

The results of these calculations are shown in Tables 2.2 and 2.3 respectively, based on incremental and total installed costs for different tax regulations (present tax regulations, 50% investment tax credit and 25% tax rate) and financing (50%) for steam turbine topping plant supplying 250,000 lbs/hr of 150 psig steam and installation between 1975–1985 for all regions. Figure 2.18 presents the calculations for the percent of companies willing to invest parametrically versus steam flow for industry and utility financing for installed and incremental costs for the South Atlantic Region, which correspond approximately to average U.S. costs.

From Tables 2.2 and 2.3 and Figure 2.18, we see that only a few companies in either of the three industries would invest in steam turbine topping based upon total installed costs, i.e., replacement of existing useful process steam boilers under present tax regulations. However, based upon incremental costs, i.e., for new facilities, a substantially larger number of companies in industry would be willing to invest, rising to 50% for a steam flow of 375,000 lb/hr. Changing the tax regulations and providing financing can have a significant effect on inducing industries to invest. With 50% financing or 50% investment tax credit, one-half of industry should be willing to invest at steam flows of 400,000 and 250,000 lb/hr, respectively, based upon total costs. Based on incremental costs, 50% of industry should be willing to invest in virtually the smallest size power system. Because of lower required rates of return, utilities should invest to a far greater degree than industry, and, in fact, should not require the large financial incentives required by industry. Based upon incremental costs, we see from Figure 2.18 that no incentives for utilities are required.

Gas Turbine Topping

Internal rate of return and the corresponding percent of industry and utilities willing to invest for oil-fired gas turbine topping supplying 250,000 lb/hr process steam, is shown in Table 2.4. Figure 2.19 shows the percent of industries and utilities willing to invest as a function of steam flow rate for the South Atlantic Region. From this table and figure we see that, except for the Mountain Region, internal rates of return are greater than 15% and thus a substantial portion of industry

TABLE 2.2: STEAM TURBINE TOPPING—RATE OF RETURN BASED ON TOTAL INSTALLED COSTS*

			. . . Internal Rate of Return, % . . .				. . Industry Willing to Invest, % . .				. . Utilities Willing to Invest, % . . .			
Region	Fuel Type	Payback Period (yr)	Present Tax Reg.	50% Investment Tax Credit	25% Tax Rate	50% Financing	Present Tax Reg.	50% Investment Tax Credit	25% Tax Rate	50% Financing	Present Tax Reg.	50% Investment Tax Credit	25% Tax Rate	50% Financing
New England	Coal	10.8	10.5	20.0	12.5	15.5	2.0	40.0	8.0	20.0	20.0	95.0	49.0	75.0
	Oil	8.8	12.5	23.0	15.5	19.5	5.0	53.0	19.0	38.0	43.0	98.0	75.0	96.0
Middle Atlantic	Coal	7.1	15.0	26.5	18.5	24.5	20.0	69.0	34.0	62.0	77.0	100.0	93.0	100.0
	Oil	6.5	16.5	16.5	20.0	27.0	25.0	74.0	41.0	69.0	84.0	100.0	95.0	100.0
East North Central	Coal	9.0	12.5	22.5	15.0	19.0	9.0	52.0	18.0	36.0	40.0	98.0	72.0	95.0
	Oil	8.0	13.5	24.5	16.5	22.0	14.0	60.0	26.0	48.0	60.0	100.0	85.0	99.0
West North Central	Coal	9.2	12.0	22.0	15.0	19.0	8.0	50.0	16.0	34.0	36.0	98.0	70.0	95.0
	Oil	8.2	13.5	24.0	16.0	21.5	13.0	58.0	24.0	46.0	55.0	100.0	84.0	98.0
South Atlantic	Coal	8.8	12.5	23.0	15.5	20.0	10.0	53.0	19.0	38.0	43.0	99.0	77.0	96.0
	Oil	8.1	13.5	24.5	16.0	21.5	13.0	58.0	24.0	46.0	58.0	100.0	83.0	98.0
East South Central	Coal	8.9	12.5	23.0	15.5	19.5	9.0	53.0	19.0	37.0	41.0	98.0	75.0	96.0
	Oil	7.6	14.0	25.5	14.5	23.0	17.0	65.0	29.0	54.0	67.0	100.0	88.0	99.0
West South Central	Coal	13.4	8.5	23.5	15.5	11.5	0.0	23.0	1.0	6.0	7.0	84.0	20.0	32.0
	Oil	19.0	5.5	13.5	6.5	6.0	0.0	12.0	0.0	0.0	0.0	57.0	1.0	0.0
Mountain	Coal	11.8	9.5	19.0	10.6	14.0	1.0	35.0	4.0	14.0	14.0	91.0	37.0	59.0
	Oil	23.0	4.0	13.5	4.5	3.0	0.0	7.0	0.0	0.0	0.0	40.0	0.0	0.0
Pacific	Coal	10.4	11.0	20.5	13.0	16.5	3.0	42.0	10.0	24.0	23.0	96.0	54.0	85.0
	Oil	14.0	8.0	14.5	9.5	11.0	0.0	24.0	1.0	4.0	5.0	81.0	18.0	22.0

*Cost of installed cost of boiler, fuel and ash handling, pollution control and turbine.

TABLE 2.3: STEAM TURBINE TOPPING—RATE OF RETURN BASED ON INCREMENTAL INSTALLED COSTS*

			. . . Internal Rate of Return, % . . .				. . . Industry Willing to Invest, % . . .				. . . Utilities Willing to Invest, % . . .			
Region	Fuel Type	Payback Period (yr)	Present Tax Reg.	50% Investment Tax Credit	25% Tax Rate	50% Financing	Present Tax Reg.	50% Investment Tax Credit	25% Tax Rate	50% Financing	Present Tax Reg.	50% Investment Tax Credit	25% Tax Rate	50% Financing
New England	Coal	5.6	18.5	31.0	22.0	31.0	33.0	81.0	53.0	78.0	94.0	100.0	99.0	100.0
	Oil	4.8	21.0	35.0	25.0	35.0	41.5	85.0	65.0	84.0	92.0	100.0	98.0	100.0
Middle Atlantic	Coal	3.7	25.0	41.0	32.0	44.0	60.0	90.0	83.0	90.0	100.0	100.0	100.0	100.0
	Oil	3.7	25.0	41.0	32.0	44.0	60.0	90.0	83.0	90.0	100.0	100.0	100.0	100.0
East North Central	Coal	4.7	21.0	35.0	25.0	36.0	44.0	86.0	66.0	85.0	97.0	100.0	100.0	100.0
	Oil	4.5	22.0	36.0	28.0	37.0	47.0	87.0	70.0	86.0	98.0	100.0	100.0	100.0
West North Central	Coal	6.4	16.5	28.5	20.0	27.0	26.0	75.0	44.0	70.0	87.0	100.0	96.0	100.0
	Oil	4.5	22.5	36.0	28.0	37.0	47.0	87.0	70.0	86.0	98.0	100.0	100.0	100.0
South Atlantic	Coal	4.8	21.0	35.0	25.0	35.0	44.0	85.0	65.0	84.0	97.0	100.0	100.0	100.0
	Oil	4.4	21.5	35.5	26.0	38.0	49.0	87.0	72.0	86.0	98.0	100.0	100.0	100.0
East South Central	Coal	4.7	21.0	35.0	25.0	36.0	44.0	86.0	66.0	85.0	97.0	100.0	100.0	100.0
	Oil	4.3	22.5	37.5	28.0	36.0	51.0	87.0	73.0	87.0	99.0	100.0	100.0	100.0
West South Central	Coal	7.0	15.5	27.5	19.0	25.0	21.0	70.0	35.0	63.0	80.0	100.0	93.0	100.0
	Oil	10.6	10.5	20.0	12.5	16.0	3.0	41.0	9.0	22.0	21.0	95.0	52.0	83.0
Mountain	Coal	6.2	17.0	29.0	21.0	28.0	28.0	77.0	45.0	72.0	90.0	100.0	97.0	100.0
	Oil	13.0	9.0	17.5	10.5	12.0	0.0	28.0	2.0	7.0	8.0	86.0	22.0	38.0
Pacific	Coal	5.5	18.5	31.5	22.5	31.5	35.0	81.0	55.0	79.0	94.0	100.0	99.0	100.0
	Oil	7.8	14.0	26.0	17.0	22.5	15.0	62.0	27.0	52.0	64.0	100.0	86.0	99.0

*Cost of steam turbine topping system less process steam boiler.

FIGURE 2.18: PERCENT OF COMPANIES WILLING TO INVEST vs STEAM FLOW FOR COAL-FIRED 1,250 PSIG STEAM TURBINE TOPPING FOR INSTALLATION 1975–1985

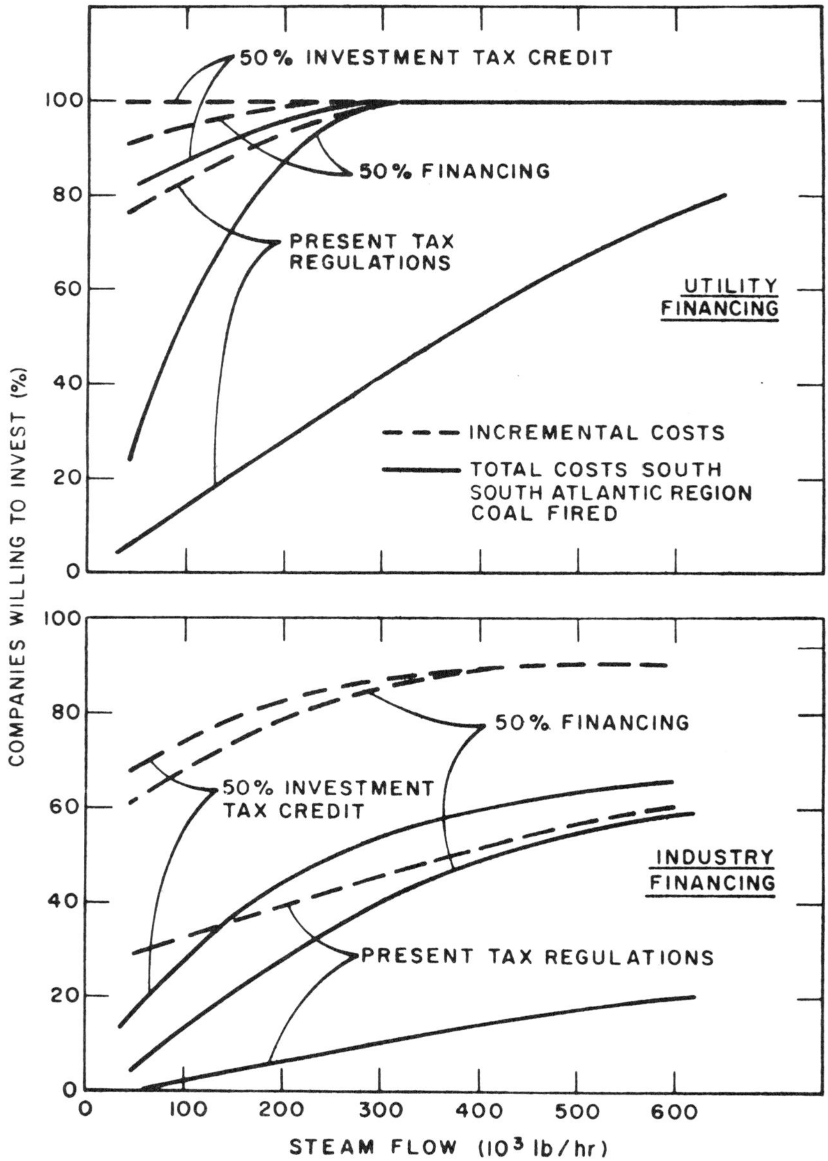

TABLE 2.4: OIL-FIRED GAS TURBINE TOPPING

Region	Type* of Analysis	Payback Period (yr)	. . . Internal Rate of Return, % . . . Present Tax Reg.	50% Investment Tax Credit	25% Tax Rate	50% Financing	. . . Industry Willing to Invest, % . . . Present Tax Reg.	50% Investment Tax Credit	25% Tax Rate	50% Financing	. . . Utilities Willing to Invest, % . . . Present Tax Reg.	50% Investment Tax Credit	25% Tax Rate	50% Financing
New England	T	4.3	22.0	37.0	28.0	38.0	51.0	88.0	74.0	88.0	100.0	100.0	100.0	100.0
	I	2.8	30.0	51.0	39.0	55.0	76.0	94.0	91.0	94.0	100.0	100.0	100.0	100.0
Middle Atlantic	T	2.65	32.0	53.0	42.0	58.0	78.0	95.0	92.0	95.0	100.0	100.0	100.0	100.0
	I	1.6	48.0	80.0	55.0	80.0	94.0	98.0	97.0	98.0	100.0	100.0	100.0	100.0
East North Central	T	3.0	29.0	48.0	37.0	52.0	72.0	94.0	90.0	94.0	100.0	100.0	100.0	100.0
	I	1.8	42.0	75.0	57.0	77.0	93.0	98.0	96.0	98.0	100.0	100.0	100.0	100.0
West North Central	T	3.4	27.0	43.0	34.0	47.0	65.0	92.0	87.0	92.0	100.0	100.0	100.0	100.0
	I	2.1	38.0	63.0	51.0	72.0	87.0	97.0	95.0	97.0	100.0	100.0	100.0	100.0
South Atlantic	T	3.2	28.0	45.0	35.0	50.0	69.0	93.0	88.0	93.0	100.0	100.0	100.0	100.0
	I	2.0	39.0	64.0	52.0	74.0	90.0	97.0	95.0	97.0	100.0	100.0	100.0	100.0
East South Central	T	2.7	32.0	53.0	42.0	57.0	77.0	95.0	92.0	95.0	100.0	100.0	100.0	100.0
	I	1.7	45.0	77.0	56.0	78.0	93.0	98.0	96.0	98.0	100.0	100.0	100.0	100.0
West South Central	T	6.0	17.0	30.0	22.0	29.0	30.0	74.0	50.0	78.0	91.0	100.0	100.0	98.0
	I	4.0	23.0	39.0	27.0	41.0	55.0	89.0	70.0	89.0	100.0	100.0	100.0	100.0
Mountain	T	18.0	6.0	14.0	–	7.0	0.0	1.3	0.0	0.0	0.0	4.0	0.0	2.0
	I	12.0	10.0	29.0	–	14.0	0.0	33.0	15.0	13.0	14.0	36.0	60.0	54.0
Pacific	T	7.2	15.0	27.0	18.0	25.0	19.0	69.0	34.0	59.0	76.0	100.0	100.0	100.0
	I	4.4	22.0	37.0	27.0	38.0	49.0	88.0	72.0	87.0	98.0	100.0	100.0	100.0

*T is total, based on installed costs of gas turbine and waste heat boiler; and I is incremental, based on installed costs of gas turbine and waste heat boiler less process steam boiler.

FIGURE 2.19: PERCENT OF COMPANIES WILLING TO INVEST vs STEAM FLOW FOR OIL-FIRED GAS TURBINE TOPPING FOR INSTALLATION 1975–1985 SOUTH ATLANTIC REGION

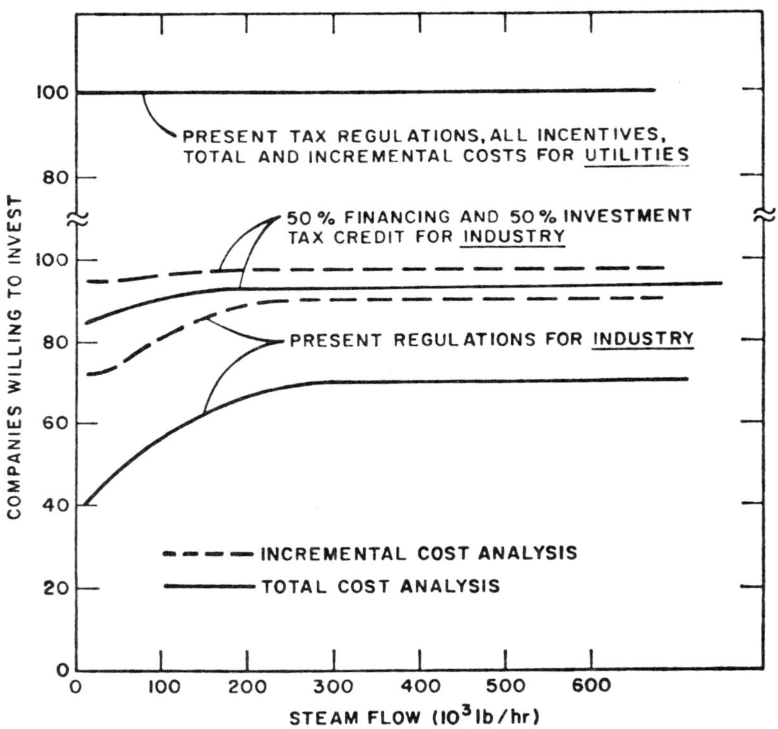

and a greater portion of utilities should be willing to invest in gas turbine topping, even for small size plants. Gas turbine topping provides greater rates of return than steam turbine topping in spite of higher costs for fuel, operation and maintenance, since capital costs are substantially lower (Figures 2.6 and 2.10). The economics of gas-fired gas turbines, based upon the fuel prices shown in Table 2.1, are even better than for oil-fired systems. The requirements for incentives for enhancing implementation of gas turbine topping are minimal for industry ownership. The above discussion should be considered in light of the fact that state-of-the-art gas turbines currently burn oil or natural gas; both of these fuels are, and will continue to be, in short supply.

Diesel Topping

Table 2.5 and Figure 2.20 repeat the calculations discussed in the previous sections for steam turbine topping and gas turbine topping. Rates of return for diesel topping are not as high as those for gas turbine topping, but are higher than those for steam turbine topping, with rates of return generally higher than 15% with the present tax regulation. The exceptions are: the Mountain Region where, in fact, diesel topping fuel plus operating and maintenance costs are greater than the cost of utility electricity; and the West South Central Region. Thus, as with gas turbine topping, a substantial portion of industry and an even greater portion of utilities should be willing to invest in diesel topping, and the requirements for incentives to enhance implementation are minimal. Like gas turbines, diesels currently burn oil or natural gas, fuels in sort supply.

Comparison of the Economics of Topping Systems

Table 2.6 shows a comparison for the economics of steam, gas turbine and diesel topping systems for a 250,000 lb/hr steam flow and for the South Atlantic Region. It may be seen from this table (as noted previously) that gas turbine topping provides a higher rate of return than either steam turbine or diesel topping, with diesel topping providing a higher rate of return than steam turbine topping. It is significant to note that the differences in rate of return, based upon a total cost analysis, are substantial, reducing considerably when based upon an incremental cost analysis.

Steam and Organic Rankine Bottoming Systems (ORCS)

Internal rates of return and the corresponding percent of industry and

TABLE 2.5: OIL-FIRED DIESEL TOPPING

			. . . Internal Rate of Return, % . . .				. . Industry Willing to Invest, % . .				. . .Utilities Willing to Invest, % . . .			
Region	Type* of Analysis	Payback Period (yr)	Present Tax Reg.	50% Investment Tax Credit	25% Tax Rate	50% Financing	Present Tax Reg.	50% Investment Tax Credit	25% Tax Rate	50% Financing	Present Tax Reg.	50% Investment Tax Credit	25% Tax Rate	50% Financing
New England	T	6.6	16.0	28.0	20.0	27.0	24.0	73.0	40.0	68.0	85.0	100.0	95.0	100.0
	I	5.3	19.0	32.0	24.0	32.0	37.0	82.0	57.0	81.0	95.0	100.0	99.0	100.0
Middle Atlantic	T	3.9	34.0	40.0	31.0	42.0	57.0	90.0	81.0	90.0	100.0	100.0	100.0	100.0
	I	3.2	28.0	46.0	36.0	50.0	69.0	93.0	89.0	93.0	100.0	100.0	100.0	100.0
East North Central	T	4.8	20.0	34.0	25.0	35.0	44.0	86.0	65.0	84.0	92.0	100.0	98.0	100.0
	I	3.9	24.0	40.0	31.0	42.0	57.0	90.0	81.0	90.0	100.0	100.0	100.0	100.0
West North Central	T	4.8	20.0	34.0	25.0	35.0	44.0	86.0	65.0	84.0	92.0	100.0	98.0	100.0
	I	4.0	24.0	39.0	30.0	41.0	55.0	89.0	80.0	89.0	100.0	100.0	100.0	100.0
South Atlantic	T	4.8	20.0	34.0	25.0	35.0	44.0	86.0	65.0	84.0	92.0	100.0	98.0	100.0
	I	4.0	24.0	39.0	30.0	41.0	55.0	89.0	80.0	89.0	100.0	100.0	100.0	100.0
East South Central	T	3.8	24.0	40.0	31.0	43.0	58.0	89.0	80.0	89.0	100.0	100.0	100.0	100.0
	I	3.1	28.0	47.0	37.0	51.0	70.0	94.0	90.0	94.0	100.0	100.0	100.0	100.0
West South Central	T	9.1	12.0	23.0	15.0	19.0	8.0	52.0	17.0	35.0	39.0	99.0	70.0	95.0
	I	7.7	14.0	26.0	17.0	23.0	16.0	64.0	29.0	52.0	66.0	100.0	88.0	99.0
Mountain	T	–	–	–	–	–	–	–	–	–	–	–	–	–
	I	–	–	–	–	–	–	–	–	–	–	–	–	–
Pacific	T	5.0	20.0	33.0	25.0	34.0	40.0	84.0	61.0	83.0	91.0	100.0	100.0	100.0
	I	4.0	23.0	39.0	30.0	40.0	55.0	89.0	80.0	89.0	96.0	100.0	100.0	100.0

*T is total, based on installed costs of diesel and waste heat boiler; and I is incremental, based on installed costs of diesel and waste heat boiler less process steam boiler.

FIGURE 2.20: PERCENT OF COMPANIES WILLING TO INVEST vs STEAM FLOW FOR OIL-FIRED DIESEL TOPPING FOR INSTALLATION 1975–1985 SOUTH ATLANTIC REGION

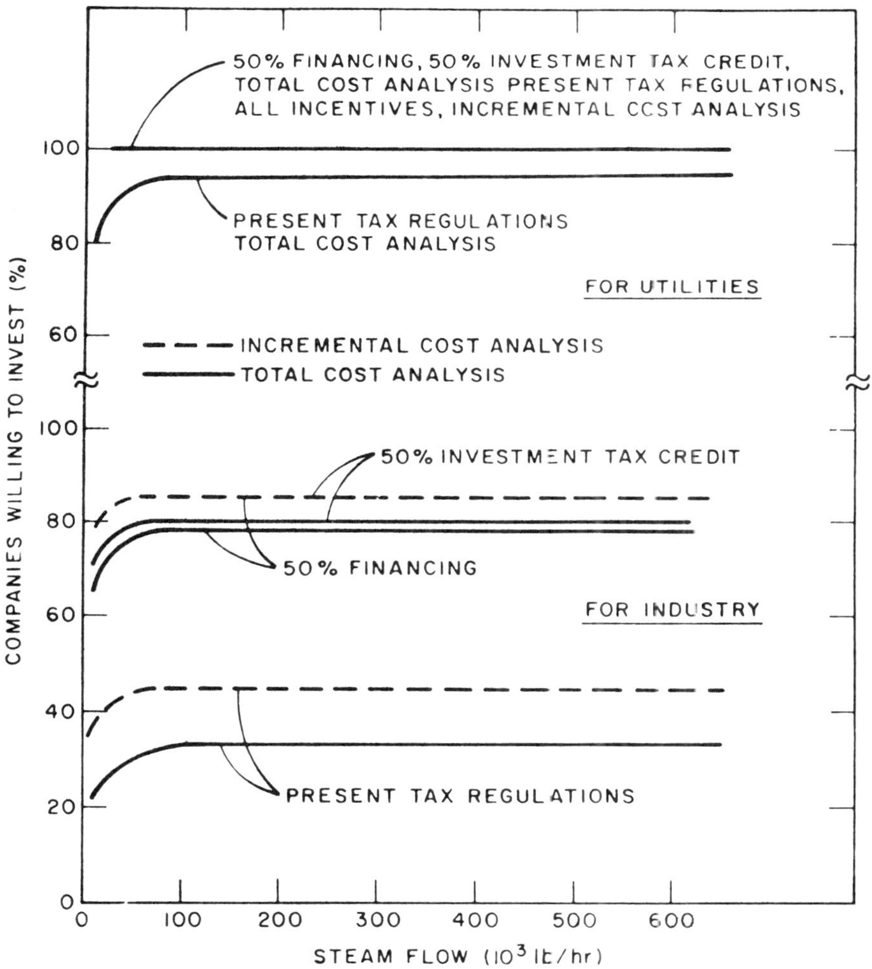

TABLE 2.6: COMPARISON OF ECONOMICS FOR TOPPING SYSTEMS

Type of Topping System	Type* of Analysis	Fuel	Payback Period Years	Internal Rate of Return	Industry Willing to Invest	Utilities Willing to Invest
				Percent		
Steam Turbine	Total	Coal	8.8	17.5	10.0	43.0
		Oil	8.1	13.5	13.0	58.0
Steam Turbine	Incre.	Coal	4.8	21.0	44.0	97.0
		Oil	4.4	21.5	49.0	98.0
Gas Turbine	Total	Oil	3.2	28.0	69.0	100.0
Gas Turbine	Incre.	Oil	2.0	39.0	90.0	100.0
Diesel	Total	Oil	4.8	20.0	44.0	92.0
Diesel	Incre.	Oil	4.0	24.0	55.0	100.0

*Total - Based on All Components of Topping System
Incremental - Based on All Components of Topping System Less Process Steam Boiler

TABLE 2.7: STEAM AND ORCS BOTTOMING

Region	Type* of Analysis	Payback Period (yr)	. . . Internal Rate of Return, % . . . Present Tax Reg.	50% Investment Tax Credit	25% Tax Rate	50% Financing	. . Industry Willing to Invest, % . . Present Tax Reg.	50% Investment Tax Credit	25% Tax Rate	50% Financing	. . .Utilities Willing to Invest, % . . . Present Tax Reg.	50% Investment Tax Credit	25% Tax Rate	50% Financing
New England	Total	4.82	21.0	35.0	25.0	35.0	42.0	85.0	65.0	84.0	97.0	100.0	100.0	100.0
Middle Atlantic	Total	3.97	23.0	39.0	30.0	41.0	55.0	89.0	80.0	89.0	99.0	100.0	100.0	100.0
East North Central	Total	4.85	21.0	35.0	25.0	35.0	42.0	85.0	65.0	84.0	97.0	100.0	100.0	100.0
West North Central	Total	5.49	19.0	32.0	23.0	32.0	35.0	81.0	55.0	79.0	94.0	100.0	99.0	100.0
South Atlantic	Total	4.51	22.0	36.0	27.0	37.0	48.0	87.0	70.0	86.0	98.0	100.0	100.0	100.0
East South Central	Total	4.47	22.0	36.0	27.0	37.0	48.0	87.0	70.0	86.0	98.0	100.0	100.0	100.0
West South Central	Total	6.40	17.0	29.0	20.0	28.0	31.0	75.0	42.0	70.0	87.0	100.0	96.0	100.0
Mountain	Total	5.69	13.0	31.0	22.0	31.0	33.0	80.0	52.0	77.0	93.0	100.0	98.0	100.0
Pacific	Total	4.83	21.0	35.0	25.0	35.0	42.0	85.0	65.0	84.0	97.0	100.0	100.0	100.0

FIGURE 2.21: PERCENT OF COMPANIES WILLING TO INVEST vs WASTE HEAT FOR ORCS AND STEAM BOTTOMING

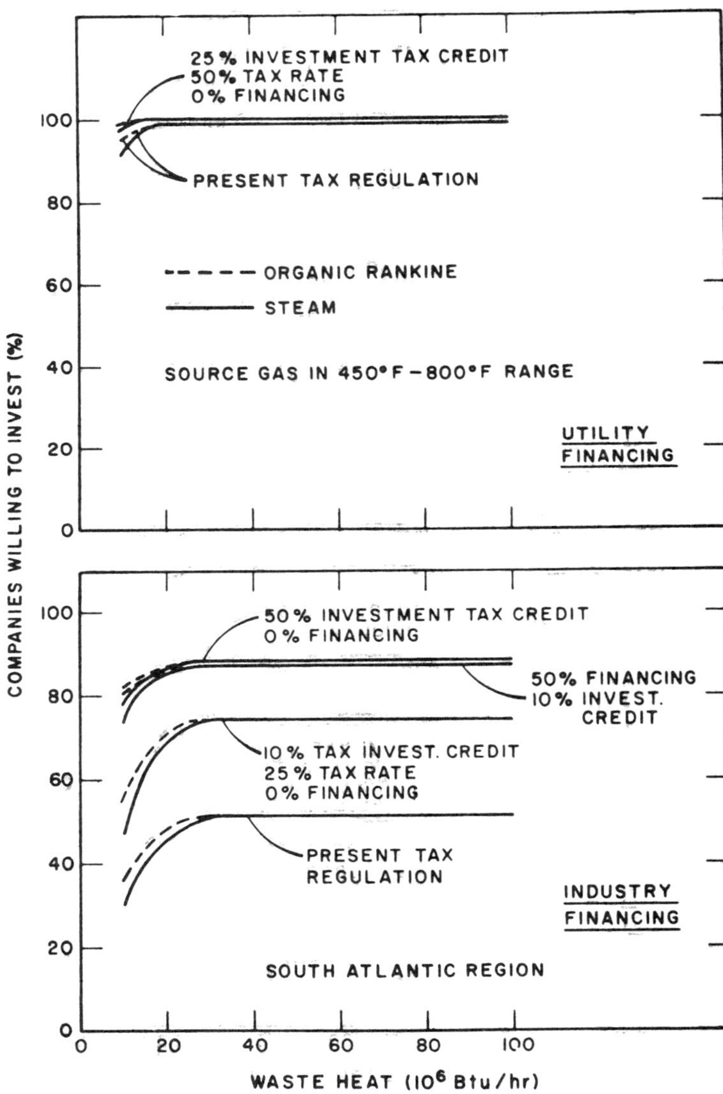

utilities willing to invest in steam and organic Rankine bottoming systems for waste heat rejection, rates in excess of 40 x 10^6 Btu/hr, are shown in Table 2.7. Figure 2.21 shows the percent of industries and utilities willing to invest as a function of waste heat rejection rate for the South Atlantic Region. Under present tax regulations, the internal rate of return for bottoming systems in the South Atlantic Region is about the same as that of the diesel topping system. It should be pointed out that, whereas the economics are essentially the same for both types of bottoming systems, the organic Rankine system develops substantially more power than the steam bottoming system.

COSTS OF INPLANT GENERATED ELECTRICITY

The fuel component of the projected 1985 bus bar costs were calculated using the appropriate power chargeable heat rates (Figure 2.22) and the projected 1985 regional fuel costs (Table 2.1).

FIGURE 2.22: HEAT RATES vs STEAM PRESSURE

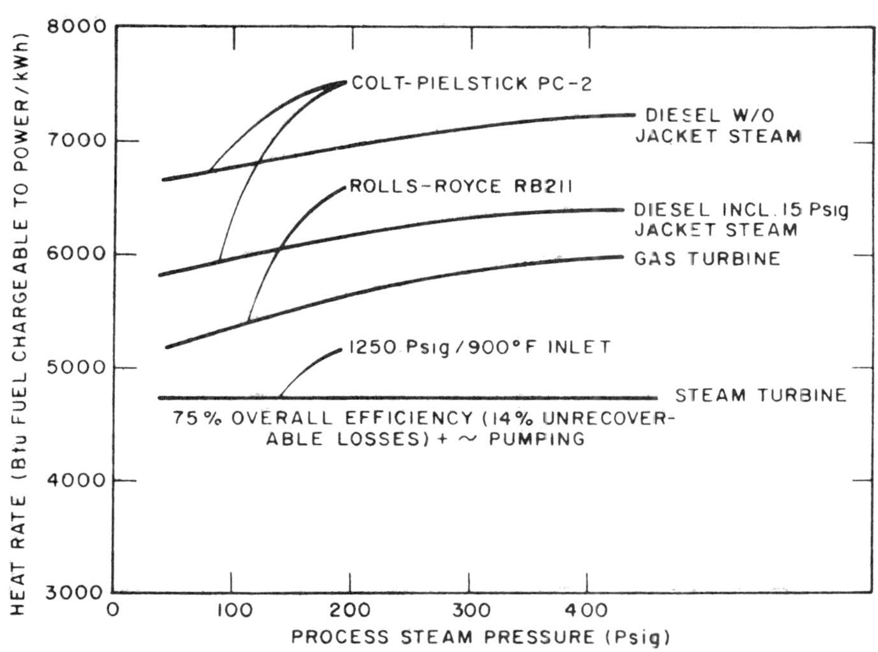

TABLE 2.8: 1985 PRICE FORECASTS OF INPLANT GENERATED vs CENTRAL UTILITY GENERATED ELECTRICITY (¢/kWh)

	Topping of Process Steam								
	Steam Turbine			Gas Turbine			Diesel		
Region	Bus Bar	Average Cost	Industrial Cost	Bus Bar	Average Cost	Industrial Cost	Bus Bar	Average Cost	Industrial Cost
New England	3.84	6.14	4.61	3.41	5.46	4.10	4.15	6.64	4.98
Middle Atlantic	3.08	4.93	3.70	3.25	5.20	3.90	3.95	6.32	4.74
East North Central	2.94	4.70	3.53	2.63	4.21	3.16	3.20	5.12	3.84
West North Central	2.91	4.66	3.50	2.39	3.82	2.87	2.91	4.66	3.50
South Atlantic	3.05	4.88	3.66	2.97	4.75	3.56	3.61	5.78	4.34
East South Central	2.89	4.62	3.47	2.31	3.70	2.78	2.81	4.50	3.38
West South Central	2.72	4.35	3.26	2.82	4.51	3.38	3.43	5.49	4.12
Mountain	2.76	4.42	3.32	4.01	6.42	4.82	4.87	7.79	5.84
Pacific	1.84	2.94	2.21	2.75	4.40	3.30	3.34	5.34	4.01

	Bottoming of Waste Heat Steam and Organic					
Region	Bus Bar	Average Cost	Industrial Cost	Industrial* Cost of Fossil Fuel Utility Generated Power	Average* Cost of Fossil Fuel Utility Generated Power	Industrial* Cost Averaged Over All Utility Generated Power Mix
New England	2.73	4.37	3.28	7.21	9.60	4.15
Middle Atlantic	2.73	4.37	3.28	5.66	7.53	5.5
East North Central	2.73	4.37	3.28	4.11	5.48	4.9
West North Central	2.73	4.37	3.28	3.16	4.21	4.4
South Atlantic	2.73	4.37	3.28	3.91	5.20	5.2
East South Central	2.73	4.37	3.28	3.30	4.40	5.7
West South Central	2.73	4.37	3.28	8.82	11.74	4.2
Mountain	2.73	4.37	3.28	2.68	3.57	4.7
Pacific	2.73	4.37	3.28	4.28	5.69	5.5

*From Mathematica Inc.

FIGURE 2.23: BUS BAR COSTS OF INPLANT GENERATED POWER USING STEAM FLOW RATE vs STEAM FLOW RATE FOR SOUTH ATLANTIC REGION

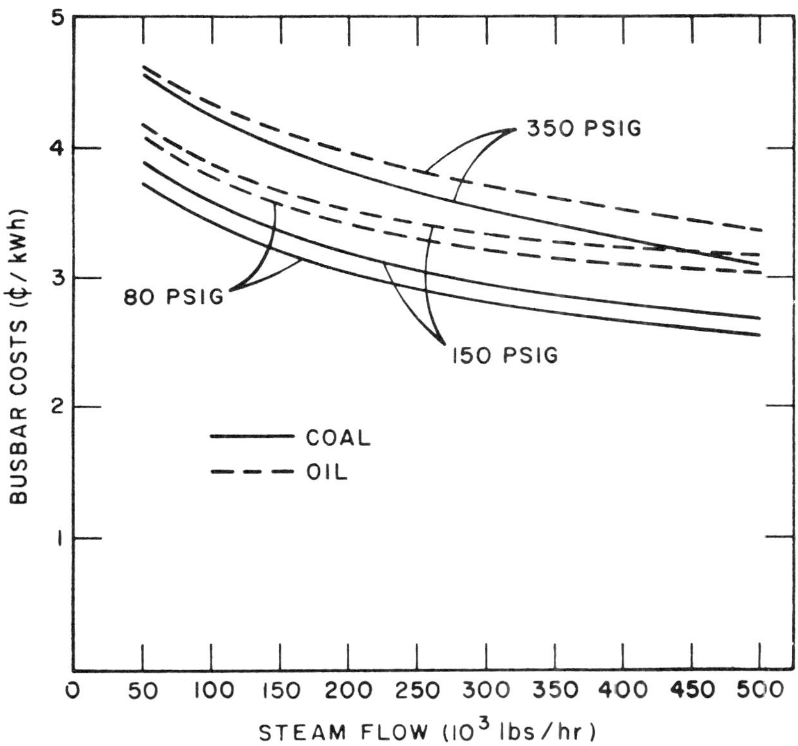

FIGURE 2.24: BUS BAR COSTS OF INPLANT GENERATED POWER vs STEAM FLOW RATE SOUTH ATLANTIC REGION

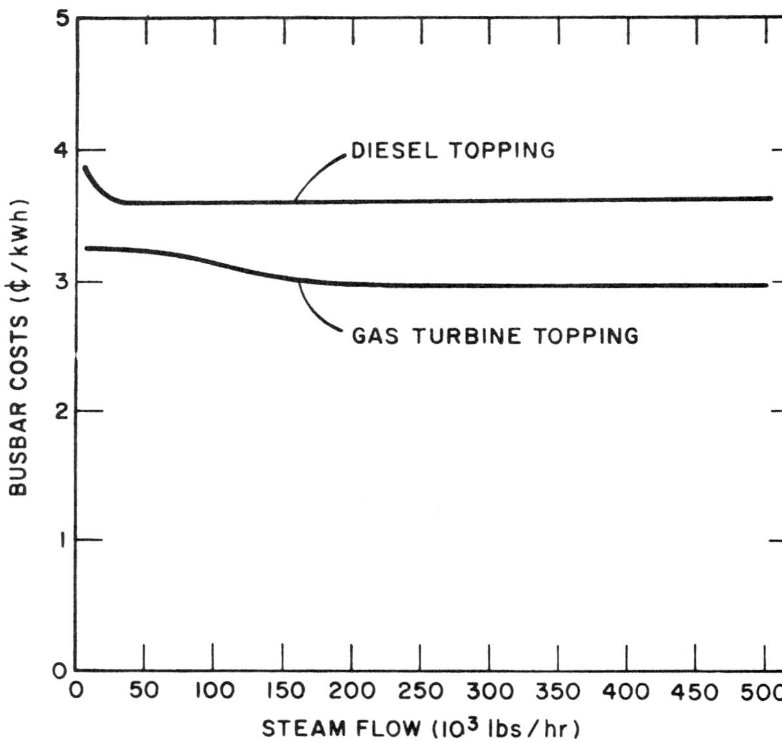

FIGURE 2.25: BUS BAR COSTS OF INPLANT GENERATED POWER vs WASTE HEAT SOUTH ATLANTIC REGION

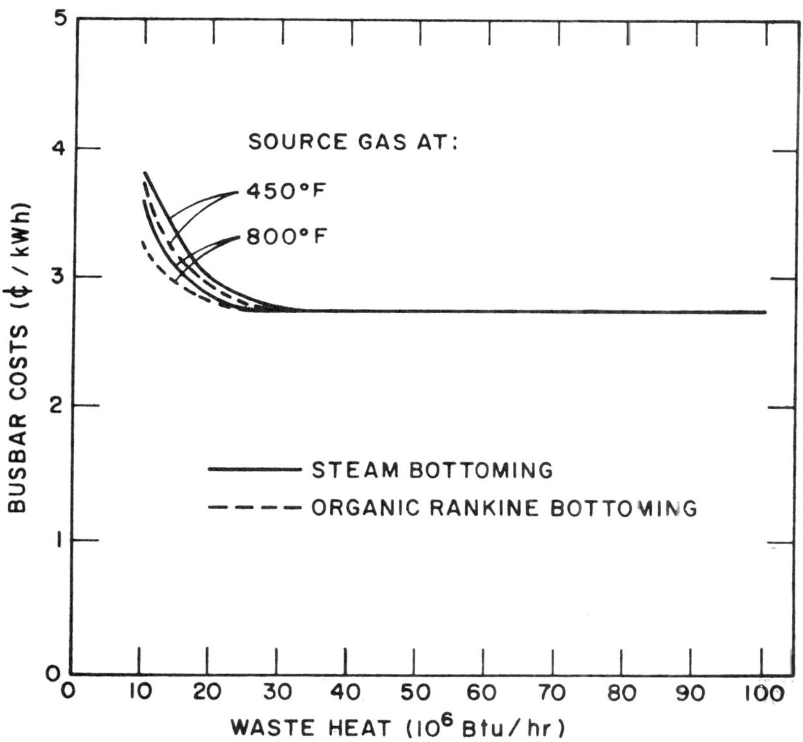

The consumer price index was then used to evaluate the 1985 operation and maintenance costs. A capital charge of 15% was assumed in determining the capital costs which were adjusted with the aid of the construction price index. The sum of these three components represents the bus bar costs which are shown in Table 2.8 for the various types of steam topping and waste heat bottoming installations as a function of region. The variations of bus bar costs with steam flow and with waste heat for the South Atlantic Region are plotted in Figures 2.23 to 2.25. As expected, the bus bar costs of waste heat generated electricity is lower than the topping systems because no fuel is involved.

The total cost of electricity averaged over commercial-residential and industrial customers (Table 2.8) was obtained by multiplying the bus bar costs by a factor of 1.6 which is somewhat smaller than the one currently in use which is 1.7 to account for the lower transmission and distribution costs to industrial users of inplant generated power. The total cost to industrial users was determined by multiplying the average cost by 0.75.

A comparison between average/industrial costs of inplant generated electricity and average/industrial costs of fossil fueled utility generated power shows that inplant power generation by steam turbine topping is cheaper in all but the West North Central, the East South Central, and the Mountain Regions. Diesel topping is cheaper in all but the West North Central, the South Atlantic, the East South Central, and the Mountain Regions. Gas turbine topping and waste heat bottoming are cheaper in all but the Mountain Region. The situation differs with respect to industrial costs of utility generated electricity averaged over all power sources (including hydroelectric, geothermal, and fission reactors). Steam turbine topping is cheaper in all but the New England Region, gas turbine topping is cheaper in all but the Mountain Region, and diesel topping is cheaper in all but the New England and Mountain Regions. Predictably, waste heat bottoming is cheaper in all regions.

In general the lowest inplant generated costs of electricity are by gas turbine topping in the East and West North Central Regions, by steam topping in the West South Central and Pacific Regions, and by waste heat bottoming in all other regions.

SUMMARY

Inplant generation of electricity in industry can be an extremely important energy conservation strategy and will help to increase the ratio of Gross National Product to energy consumption per capita. This conclusion is supported by the potential in the chemical, petroleum refining and paper and pulp industries studied in this report.

In this study only power systems which generate electric power in conjunction with process steam were considered. Thus utility type systems, such as condensing steam turbines, gas turbines and diesels without waste heat boilers, and combined gas turbine and steam turbine power plants were excluded.

Potentials for Inplant Electric Power Generation

The thermodynamic potential for inplant generation and for export of power depends strongly on the type of system. In most regions of the country, there is a good balance between the power that can be produced from a combination of steam turbine topping and bottoming of waste heat and the electricity required by industry for self-sufficiency. The primary exception is the South West Central area of the U.S. where more than 15% of the utility generated electricity in 1974 could be exported as a result of the large concentration of chemical plants and petroleum refineries in this region. Nationally, with gas turbines and diesel topping, the industries studied could export the equivalent of 30 and 68%, respectively, of utility generated power in 1974. In the South West Central Region, more power than required by utilities could be generated inplant by these industries with gas turbine and diesel topping. The applicability of gas turbine and diesel power systems is generally limited regionally, however, by the ability of these systems to operate presently only with gaseous and liquid fuels.

These potentials for inplant generation can be compared to industrial production of electricity in 1974 in the three industries amounting to about 30% of requirements and about 3.4% of total utility generation with negligible export of power. Thus, only a fraction of the potential for inplant generation in the chemical, petroleum refining and paper and pulp industries is currently being realized. It is noteworthy that these three industries currently produce about two-thirds of the electricity generated inplant by all industry in the U.S., yet use only about 25% of the process steam. Other industries, therefore, generate even less of their potential than these three industries. In fact, inplant electric power generation has been on the decline in the U.S. in recent years, declining from 15% of the nation's total electric power in 1950 to about 5% in 1973. The potential for inplant generation in the U.S. can be verified by comparison to West Germany, an industrialized society similar in industrial output to the U.S., where about 29% of total electric power is produced by industry.

The fuel savings based upon the maximum thermodynamic potential for inplant generation in 1974 is the fuel equivalent of about 550,000, 1,500,000 and 2,200,000 bbl oil/day for steam turbine, gas turbine and diesel topping, respectively.

If the present trends for inplant generation were to continue, it is estimated that the percent of self-generated power used by industry would remain essentially at today's levels and decline as a percentage of total utility generation (as a result of the greater growth rate of utility consumption), despite higher fuel costs and rates of returns that are attractive compared to average industry returns. The primary reasons for this can be attributed to the basic characteristics of our industrial society.

Industry operates on a competitive basis. Profits must be earned that are large enough to attract additional capital for expansion. Larger dollar volumes of capital can be earned from product oriented investments than from inplant generation or other energy conservation strategies. To ensure earnings in their primary product oriented business areas, industry invests heavily in product oriented ventures leaving little capital for ancillary investments such as inplant generation. The capital for product oriented ventures is obtained from retained earnings, increasing equity, or increasing debt. Many industries have been plagued by a severe shortage of equity capital since the late 1960s, as a result of depressed stock market conditions and poor profits, requiring them to finance investments by increasing debt.

Usually, a company has expended its returned earnings and does not want to or cannot increase equity further (as a result of a low book value to common equity ratio) to finance inplant generation or other fuel conserving ancillary investments. Industry must thus increase debt further to obtain funds for these projects. Again, it is limited in the amounts of funds it can obtain because of an already high debt to equity ratio resulting from past borrowing for product oriented ventures. In addition, the cost of borrowed funds increases as the debt to equity ratio of the company increases. The net result is that funds used to finance ancillary investments, such as inplant generation, are more expensive than funds used for financing product oriented ventures. Thus, historically, inplant generation of electricity has been considered by industry, except in isolated cases, as an ancillary investment requiring higher rates of return since it is not normally associated with the profitability or expansion of its primary market.

Power generation is the primary product of utilities. They should be anxious to invest in inplant generation, since in many regions and situations inplant generation could result in a return equal to or greater than that obtainable from central station power plants. In fact, we estimate that utilities could economically generate 70 to 80% of the thermodynamic maximum potential for inplant generation by 1985. Based upon our discussions with utilities, we find that utilities are mostly negative to inplant generation with either industry or their own ownership of the facilities. They understandably want to maintain their place in the industrial market for electrical power and are reluctant to compete with industry owned inplant generation systems. Furthermore, present regulatory postures of the Federal Power Commission and the State Public Utility Commissions and existing rate structures, based on a fair and just return on the valuation of utility property and distributed costs, are not conducive to utility investment in inplant generation.

Economic Incentives and Advantages

We have studied several economic incentives for enhancing inplant generation, including investment tax credits and lower taxes on investment for inplant generation and financing at current interest rates. We favor government guaranteed loans perhaps up to a maximum of 50% over a 20-year period as an economical incentive to industry and leasing companies to induce them into implementing increased inplant generation although we recognize that in some segments of industry tax incentives such as investment tax credits can also significantly enhance adaption of inplant electric power generation. This recommendation is based on the following:

1) Our observation that 50% financing will tend to make industry invest in inplant generations systems which are more attractive from a national viewpoint than utility power while not substantially increasing investment in those systems for which the reverse is true.
2) Federal Government guaranteed loans offered at market rates would probably be viewed differently by the investment community than ordinary debt and could possibly be treated as an off-balance sheet entry, allowing industry to borrow funds for its regular product oriented investments.

3) The federally guaranteed loans would be especially effective if they were made available to leasing companies. The debt for eligible equipment would then not be recorded as additional debt on the process company's financial statement. It would, of course, appear on the balance sheet of the leasing company. However, it is typical of leasing companies that they are able to establish credit with numerous financial institutions, whereas an individual company normally relies on a single banking relationship.
4) Loans are judged to interfere with the free market least of all of the incentives studied. Pressure on money markets will be reduced as a result of the lower total investment required, from fuel supply to end user, for inplant generation compared to separate utility and industrial process steam systems while federal tax revenues will not be affected.

We estimate based upon our economic analysis that with federally guaranteed loans to industry and leasing companies with up to 50% of the required investment, about 60 to 70% of the potential for inplant generation could be realized, resulting in substantially more generation of power than today at costs less than or competitive with utility power. With gas turbine topping, export of power by the chemical, petroleum refining, and paper and pulp industries could amount to about 13% of the electricity generated by utilities in 1985 increasing to 24% for diesel topping. With steam turbine topping central station power would still be required in these three industries, but inplant generated power would increase to 67% of requirements, up from 30% in 1974–1975. Fuel savings in 1985 could amount to the equivalent of about 500,000, 1,600,000 and 1,900,000 bbl oil/day for steam turbine, gas turbine and diesel topping systems, respectively. Table 2.9 summarizes the economic potentials for inplant generation with and without federally guaranteed loans and industrial or leasing and utility ownership of facilities in 1985. Similar amounts of power and fuel savings can be attained with a program of 50% investment tax credits to industry for inplant electric power generation.

The optimum business arrangement for ownership of inplant generation facilities will often be predicated on local considerations and attitudes. Thus, the climate should also be improved for utility ownership even though the economics of inplant generation appears comparable to or better than central station power. In addition to elimination of regulatory constraints and changes in rate structure, we suggest a moderate increase in the investment tax credit from 10 to 12% and/or

TABLE 2.9: SUMMARY OF MAXIMUM THERMODYNAMIC AND ECONOMIC POTENTIALS AND FUEL SAVINGS FOR INPLANT GENERATION OF ELECTRICAL POWER IN 1985

	Economic Potential for Inplant Generation			Economic Potential for Export of Power			Fuel Savings		
	Steam Turbine	Gas Turbine	Diesel	Steam Turbine	Gas Turbine	Diesel	Steam Turbine	Gas Turbine	Diesel
	Percent of Utility Generation						Bbl/Oil Equivalent/Day		
1.0 Industry or leasing company ownership									
Without economic incentives	4.0	15.0	18.0	(5.0)	6.0	10.0	270,000	1,100,000	1,000,000
With federally guaranteed loans (50% of investment)	6.0	22.0	33.0	(3.0)	13.0	24.0	500,000	1,600,000	1,900,000
2.0 Utility ownership									
Without economic incentives	7.0	23.0	38.0	(2.0)	14.0	29.0	600,000	1,800,000	2,100,000
With federally guaranteed loans (50% of investment)	8.0	24.0	44.0	(1.0)	15.0	35.0	760,000	1,900,000	2,500,000
3.0 Maximum thermodynamic potential	10.0	35.0	70.0	1.0	25.0	60.0	800,000	2,200,000	3,300,000

Note: ( ) is purchase of power.

federally guaranteed loans up to 10% of investment requirements. These economic incentives will not materially change the economics of utility ownership of inplant generation facilities, but will serve to induce utilities to broaden attitudes and seek opportunities and rule changes for inplant generation.

In addition to conserving fuel, inplant generation could also conserve capital. Inplant generation, in most cases, will require investments from fuel supply to customer and user comparable to or less than that required for central station power. Capital investment for power plant alone could be reduced from $35 billion to $9 to $12 billion with gas turbine or diesel topping supplying power for industry self-sufficiency. Greater savings are also obtainable if diesel and gas turbine power is exported. Savings in investment for power plants with steam turbine topping are significant but not as substantial as with gas turbines or diesels. Steam turbine power plants can burn coal while gas turbines and diesels operate on liquid or gaseous fuels.

Barriers to Implementation

Technical and operational barriers to increased inplant generation are not regarded as forbidding. Technologies required are available and in use in industry. Fuel consumption patterns would not be substantially different for inplant generation based on steam turbine topping than it would be for central station boiler since both can be coal-fired. Substantial implementation of gas turbines and diesels could be severely limited by their ability to burn only clean liquid and gaseous fuels, although development work is under way for operation with coal or coal-derived fuels. However, in the Southwest Central Region where large amounts of natural gas are currently used for electrical generation, gas turbines could be used for inplant electrical generation without substantially changing regional fuel patterns even in 1985. Differences in utility demand and industry electrical supply schedules could limit implementation of inplant generation for export to a base load strategy in some regions for gas turbine and diesel topping.

The barriers resulting from regulatory and rate structure constraints could be more formidable than others as a result of the long lead times required for hearings and litigations. Large-scale use of inplant generation would most likely trigger many questions on the jurisdiction of the Federal Power Commission and the State Public Utility Commission. Questions concerning mode of ownership and financing of inplant generation (especially the question of treating government guaranteed loans as off-balance sheet entries) could also involve the Security and Exchange Commission as they relate to accounting and stock issuance practices. Implementation of inplant generation could be impeded by low fuel prices and curtailment of natural gas for use in gas turbines for electrical generation. These policies should be reassessed to determine their effect on inplant generation.

Implementation of Inplant Electrical Power Generation

We suggest that a multifaceted program to enhance implementation of inplant generation be initiated and coordinated by the energy agencies of the federal government. This program should consist of advanced hardware development in coal burning topping and bottoming systems for recovery of waste heat. A demonstration program for the planning, design, construction and implementation of inplant generation should be initiated in several key areas of the U.S. The demonstration program would trigger rulemaking changes and test cases in the regulatory and rate area which could lead to a more favorable climate for inplant generation. In preparation for the planning phase of the demonstration program, a rulemaking action on inplant generation should be sought and rulings from both the Federal Power Commission and State Public Utility Commissions obtained on the jurisdictional effects of different ownership options for inplant generation. Simultaneously with the advanced development and demonstration programs, the Federal Energy Agencies should sponsor programs on the following: (1) ownership and financing of inplant generation facilities and the treatment of government guaranteed loans, (2) legal and regulatory issues (including rate structuring and antitrust questions), (3) reliability and defense security of inplant power versus central station power, (4) studies in correlation of utility demand and industry electrical supply schedules (i.e., base load and/or peaking strategies), (5) information dissemination programs, and (6) practices in foreign countries.

Major Conclusions of Study

The major conclusions of this study are:

1) Inplant electric power generation can be an extremely important strategy for energy conservation in the U.S.
2) Under present economic conditions, inplant electric power generation may not increase substantially above today's levels (in terms of percent of total U.S. electrical generation) even though inplant electric power can be generated at comparable or lower

costs than fossil fuel-fired central station power in most situations. This is the result of industry's generally high required rate of return for ancillary investments such as inplant generation, utilities' generally negative attitude toward inplant generation, and the present legal and regulatory environment, which is not conducive to either industry or utility ownership of facilities for inplant generation, contributing to these attitudes.

3) A program of federally guaranteed loans to industry and leasing companies at current market interest rates and up to 50% of the required investment is preferred to enhance implementation of inplant electric power generation. Additional tax incentives such as investment tax credits could contribute significantly to inplant power generation also. Fuel savings with this program could amount to a minimum of 500,000 bbl of oil equivalent/day by 1985. Smaller incentives are recommended for utilities to induce them to invest in inplant generation.

4) The major operational barrier to export of inplant generated electric power for consumer use and greater fuel savings is that power cycles generating sufficient power for export burn gaseous and liquid fuels. Power cycles for inplant electric power generation which burn coal can economically supply about two-thirds the power required by the three industries studied with an attendant fuel savings of about 500,000 bbl of oil equivalent/day by 1985. No major technical barriers should impede implementation of inplant generation.

5) Inplant electric power generation potentially has numerous national advantages beside energy conservation. The most important of these are lower cost of electricity to the consumer and/or greater profit potential for industry and utilities, assisting in the implementation of a fuel conversion strategy to coal through the use of steam turbine topping (which would further lessen dependence on foreign oil imports), less total atmospheric pollution as a result of lower fuel consumption, and greater reliability and security of a multi-source electrical generation system versus a smaller number of larger systems.

6) A program should be initiated by a federal agency as outlined in this study consisting of hardware development, demonstration programs, and institutional and regulatory studies.

7) Inplant generation is so important to an overall U.S. fuel conservation strategy that trends in implementation should be monitored closely. Federal agencies should continually reason their strategy and make changes dictated by the rate of implementation of inplant generation.

The Potential for Cogeneration Development

Resource Planning Associates Study

The material in this chapter is taken from the report entitled *The Potential for Cogeneration Development in Six Major Industries in 1985,* prepared for the Federal Energy Administration,* Office of Conservation, by Resource Planning Associates, Inc. (RPA) and Burns and Roe, Inc. (B&R). The final report was published in December 1977. The RPA Project Director was Peter Bos, and the Project Managers were James Williams (RPA) and Jerome Morton (B&R).

*As of October 1, 1977, FEA was subsumed by the Department of Energy (DOE).

The introduction to this chapter, which explains the project and gives its scope and an overview of its conclusions, is quoted directly from the report. The rest of this chapter contains material which has been edited and summarized from the above report.

INTRODUCTION

In the past 3 years, it has become clear that actions to increase energy supplies cannot, alone, resolve the forecasted gap between domestic energy supply and demand. The United States must place much greater emphasis on energy conservation. One of the energy conservation measures receiving increased attention is the development of cogeneration plants by industries that use a significant amount of both electricity and process heat, such as the chemical, petroleum refining, paper, and steel industries.

A cogeneration plant is a power plant that produces both electricity and useful thermal energy. Because a cogeneration plant uses some of the energy that would otherwise be wasted in a separate conventional electric power plant, it can produce a given amount of electricity and thermal energy for 10 to 30% less fuel, depending on the configuration and applications involved.

Past estimates of the energy savings from increased industrial cogeneration development range from several hundred thousand barrels per day oil equivalent (bpde) to over 1 million bpde, depending on the time horizon and assumptions made. Several kinds of constraints inhibit the increased development of industrial cogeneration and, hence, the realization of these savings. They are listed below.

Technical: Because of variations in the operating requirements for electricity and process thermal energy, or because of the small amount of process energy required, some plants lack the necessary demand characteristics for a cogeneration plant.

Economic: Because of relatively low electricity prices coupled with relatively high fossil fuel prices, or because of alternative investments that have a higher return on investment (ROI), the financial attractiveness of cogeneration plants for certain industries and regions is not great enough to warrant the necessary investment.

Environmental: For plants that are technically and economically feasible, environmental standards (e.g., classification of certain air pollution districts as nondegradation or nonattainment areas) can prevent the installation and operation of a cogeneration plant.

Regulatory: Because of the uncertainties associated with the regulation of industries operating cogeneration plants, the business risk for industry investment is sometimes considered unacceptable.

Institutional: Because some companies do not consider electric power generation as part of their business, regardless of the economics, or because some utilities may not want industrial power plants connected to the electric power system in their service area, a cogeneration plant that would otherwise be technically and economically feasible may not be constructed.

FEA Study

As a result of the significant energy savings potential and the possible importance of barriers over which the federal government has partial or complete control, the Federal Energy Administration (FEA) commissioned Resource Planning Associates, Inc. (RPA), with Burns and Roe, Inc. (B&R), as a subcontractor, to assess constraints on the development of cogeneration, and to identify cost-effective government actions that could significantly increase industrial investment in cogeneration plants where cogeneration is technically and economically feasible.

There were four specific analytic boundaries set for the study:

A focus on the industrial ownership/investment decision process: Since no prior study had adequately examined the industrial investment decision-making process, a strong focus was placed on detailed analysis of all technical, economic, regulatory, environmental, and institutional factors affecting that process, recognizing industry's reluctance to consider excess power production without utility involvement.

The use of decision-making factors that actually affect the cogeneration investment decision: To reflect only real-world decision processes in our analysis, we conducted an intensive series of interviews with financial, technical, and top-line-management decision makers in key industries, and used their actual investment criteria in our economic analysis.

The use of only industrially proven technologies: In light of this parameter, we considered only coal- and oil-fired back-pressure turbines and distillate-fired gas turbine heat recovery boiler configurations, with a range of boiler and exhaust steam pressures.

The use of a time frame consistent with national energy conservation policy: Given the possible world oil supply/demand shortfall in the mid-1980s, 1985 was considered an appropriate time horizon.

Scope of the Study

Within this study scope, we carried out three main steps:

1. We identified alternative cogeneration plant configurations, and prepared engineering design and cost estimates and ROI ranges for them. B&R, which specializes in power plant design, assisted in this step. Approximately 100 companies, representing 600 potential applications of these configurations, were identified for further contact.

2. Using actual industry decision-making factors, we estimated likely cogeneration development through 1985 without government action. To this end, we interviewed executives in more than 50 companies that could use cogeneration plants, as well as representatives of 7 public utilities and 7 public utility commissions and state energy offices. These interviews were conducted in different regions of the country to ensure coverage of all major factors affecting managment decisions on cogeneration plants.

3. We identified and evaluated government actions that could be taken to increase cogeneration development significantly. Government actions that could remove or minimize the major barriers to cogeneration development were identified and their costs/benefits were examined in detail.

In estimating industrial cogeneration-development potential both without and with government action, we concentrated on those six industries that consume a significant amount of electricity and process heat: chemicals, petroleum refining, pulp and paper, steel, food processing, and textiles. Further, we focused on process steam topping applications because our initial analysis of industrial cogeneration potential indicated that 60 to 70% of the potential development is in these applications. (Direct process heat topping applications are estimated at 20 to 30% of the total potential cogeneration development, and applications using process waste heat to drive Rankine cycles are estimated at less than 10% of the potential development.) Finally, because a large proportion of energy savings was estimated to be associated with applications of greater than 5 megawatts (MW) of electrical power, we focused on applications above this size.

Although we concentrated most of our analytical effort on the process steam applications, we felt it necessary to gain some understanding of the total industrial cogeneration potential. Consequently, we estimated total industrial process energy consumption and available waste heat as a starting point in determining cogeneration potential without government action, and then developed rough estimates of the maximum technical and economic potential for electric power production, and of the energy savings from all types of cogeneration (see Table 3.1).

TABLE 3.1: ESTIMATED POTENTIAL ENERGY SAVINGS IN 1985 FROM ADDITIONAL COGENERATION DEVELOPMENT WITHOUT GOVERNMENT ACTION

Additional Cogeneration Potential*	Electric Energy Generated (billion kWh)	Electric Capacity (thousand MW)	Energy Savings	
			Trillion Btu	Thousand BPDE
Technically suitable cogeneration potential	380-740	50-100	2200-3500	1020-1650
Economically attractive cogeneration potential	100-200	10-30	500-900	250-420

*Includes cogeneration plants installed in existing and new manufacturing facilities.

Specifically, we projected the 1985 thermal energy requirements for process use in industry at 17,900 trillion Btu—13,900 trillion Btu for process steam and 4,000 trillion Btu for direct process heat (see Table 3.2). We then estimated that, in 1985, 5,525 trillion Btu—3,900 trillion Btu from topping steam and 1,625 trillion Btu from process heat—would be technically suitable for cogeneration topping plants. We also estimated the 1985 waste heat bottoming cogeneration potential in the six focus industries to be 3,540 trillion Btu, predominantly (2,760 trillion Btu) in the chemical, steel, and refining industries.

We converted this technically feasible thermal energy cogeneration potential to electric energy and energy savings. These estimates range according to the assumed mix of cogeneration plant configurations shown in Table 3.3. For instance, a back-pressure steam turbine produces roughly one-quarter as much electric energy per unit of steam as does a gas turbine/heat recovery boiler configuration, and only one-tenth the electric energy per unit of steam as a diesel engine/heat recovery boiler configuration. Using what we consider to be a feasible range of configurations for 1985, we estimated a maximum technical potential for cogeneration of 380 to 740 billion kWh of electric energy produced and energy savings of 1.0 to 1.6 million bpde (see Table 3.2).

Economic factors would further reduce the amount of potential cogeneration development by 1985, perhaps to 25% of the technically suitable cogeneration. Applying the same approach used to determine

TABLE 3.2: TOTAL INDUSTRIAL ENERGY PERSPECTIVE OF COGENERATION POTENTIAL

Industrial Fuel Consumption in 1976 (trillion Btu)

Industry Segments	Total Fuels Consumed: Purchased Fuels	Total Fuels Consumed: Process Residuals	Total Fuels Consumed: Purchased Electricity[a]	Fuels Used for Steam Generation	Fuels Used for Process Heat	Other
Six industries	**7,900**	**2,700**	**1,000**	**5,400**	4,200	2,000
Other industries	14,200	–	1,500	7,700	4,800	3,200
Total	22,100[b]	2,700	2,500	13,100[b]	9,000[b]	5,200

Industrial Process Energy Consumption (trillion Btu)

Industry Segments	Process Steam		Direct Process Heat		Process Heat Rejection	
	1976	1985	1976	1985	1976[b]	1985
Six industries	**4,100**	**5,900**	1,300	1,900	8,600[c]	10,300
Other industries	5,600	8,000	1,400	2,100	12,500[c]	15,200
Total	9,700	13,900	2,700	4,000	21,100[c]	25,500

Process Energy Technically Suitable for Additional Cogeneration[d] (trillion Btu)

Industry Segments	Process Steam		Direct Process Heat		Process Heat Rejection	
	1976	1985	1976	1985	1976	1985
Six industries	**2,200**	**3,500**	750	1,050	2,300	2,760
Other industries	280	400	460	575	650	780
Total	2,480	3,900	1,210	1,625	2,950	3,540

Technical Limit to Additional Cogeneration Development in 1985[e]

	Process Steam	Direct Process Heat	Process Heat Rejection	Total[g]
Electric energy (billion kWh)	**168-384**	173-303	35.4-51.4	**380-740**
Capacity[f] (thousand MW)	**24.0-54.8**	24.7-43.2	5.0-7.3	**50-100**
Energy savings				
(trillion Btu)	**460-1770**	830-1200	370-530	**2200-3500**
(thousand BPDE)	**450-835**	390-565	175-250	**1020-1650**

NOTES All numbers except those in boldface type have been very roughly estimated in an attempt to put the results of this study in perspective. These estimates should not be considered authoritative without further analysis.

a Estimate based on 1976 data from Edison Electric Institute.

b Estimates based on the following sources: "Energy Consumption Fuel Utilization and Conservation in Industry" prepared by Dow Chemical Company for the Environmental Protection Agency (1975), "Industrial Application Study" prepared by Drexel University for the Energy Research and Development Administration (1976). "Efficient Use of Energy," American Institute of Physics Conference Proceedings No. 25 (1975); "Annual Reviews, Inc.," (Vol. 1 1976), and "A Study of In-Plant Electric Power Generation in the Chemical, Petroleum Refining and Pulp and Paper Industries," prepared by Thermo Electron Corporation for the Federal Energy Administration (1976).

c Includes heat rejected in producing process steam and heat heat rejected in the process, and heat lost in other plant operations.

d In addition to existing cogeneration.

e See Exhibit 3 for a definition of the low and high configuration mixes used to estimate the ranges shown.

f Assumes an 80 percent annual capacity utilization.

g Numbers are rounded.

maximum technical potential, and using the range of configurations shown in Table 3.3, we estimated an economic range of cogenerated electric energy of 100 to 200 million kWh by 1985 (see Table 3.4). This corresponds to energy savings of 250 to 420 thousand bpde, which should be largely oil savings.

TABLE 3.3: FEASIBLE RANGE OF COGENERATION TECHNOLOGY MIXES FOR LOW AND HIGH ELECTRIC POWER OUTPUT

Cogeneration Application	Technology Mix (fraction) Low Electric Power Mix	High Electric Power Mix	Cogeneration Technology
Process steam	1.0	0.75	Backpressure steam turbine
	0.0	0.15	Gas turbine
	0.0	0.10	Diesel
Direct process heat	1.0	0.50	Gas turbine
	0.0	0.50	Diesel
Process waste heat	1.0	0.75	Steam Rankine (turbine)
	0.0	0.25	Organic Rankine (turbine)

Looking beyond 1985 to the 1990s and to 2000, the potential energy savings from cogeneration clearly will be higher. However, we cannot make any reasonable estimates of this future potential without further analysis. But to the extent excess power supplied to the utility system can be facilitated over the next few years, the potential savings from excess power production after 1985 could, in our opinion, easily exceed the total expected pre-1985 savings levels. Further, because new industrial manufacturing facilities will continue to replace existing facilities, and because the economics of cogeneration plants in new facilities are better than for retrofit applications, the economically attractive cogeneration (without government action) should increase significantly as a percent of the technically feasible potential.

With this understanding of total industrial cogeneration potential, in our study we sought to determine how much of the technically and economically feasible *process steam* cogeneration could be achieved by 1985, both without and with government action.

Findings

Our analysis of process steam cogeneration in the six focus industries

TABLE 3.4: COGENERATION DEVELOPMENT BY 1985 WITHOUT GOVERNMENT ACTION

Process Energy Economically Attractive for Additional Cogeneration
(trillion Btu)

Industry Segments	Process Steam		Direct Process Heat		Process Heat Rejection	
	1976	1985	1976	1985	1976	1985
Six industries	**250**	**1,160**	75	300	115	145
Other industries	25	120	23	90	16	18
Total	275	1,280	98	390	131	163

Expected Cogeneration Development in 1985 from the Economically Attractive Process Energy

	Process Steam	Direct Process Heat	Process Heat Rejection	Total
Electric energy (billion kWh)	**55-126**	42-73	1.6-2.4	**100-200**
Capacity (thousand MW)	**7.8-18.0**	6.0-10.4	0.23-0.34	**10-30**
Energy savings				
(trillion Btu)	**0.31-0.58**	0.20-0.29	0.02-0.03	**500-900**
(thousand BPDE)	**145-275**	95-135	8-12	**250-420**

NOTES All numbers except those in boldface type have been very roughly estimated in an attempt to put the results of this study in perspective. These estimates should not be considered authoritative without further analysis.

In estimating the production of electric energy from different cogeneration plant configurations, consideration must be given to the fuel type of the cogeneration plant, electric energy consumption of ancillary equipment, the conditions of the process thermal energy (e.g., pressure and flow rate), the conditions of the waste heat recovered (e.g., temperature and flow rate), and other factors. In converting from thermal energy to cogenerated electric energy, we used reasonable average conversion values to estimate reasonable ranges of electric power, capacity, and energy savings from cogeneration.
(Any specific conversion value used may not necessarily apply to a given cogeneration plant in a given industry.)
The specific conversion values used were

Process steam topping cogeneration
–43 net kWh per million Btu of process steam exhausted from noncondensing systems
–175 net kWh per million Btu of process steam generated from gas turbines with waste heat recovery boilers
–400 net kWh per million Btu of process steam generated from diesel plants with waste heat recovery boilers.

Process heat topping cogeneration
–32 net kWh per million Btu of gas turbine fuel required to produce process heat
–80 net kWh per million Btu of diesel fuel required to produce process heat.

Process waste heat bottoming cogeneration
–10 net kWh per million Btu of waste heat recovered by steam Rankine systems
–28 net kWh per million Btu of waste heat recovered by organic Rankine systems.

(summarized in Figure 3.1) indicated that government action can stimulate almost as much process steam cogeneration by 1985 as the baseline development that would normally occur, resulting in gross energy savings of 80 to 300 trillion Btu per year (or 40 to 140 thousand bpde), depending on the range of federal economic and regulatory measures implemented.

Gross energy savings are total savings, mainly of oil, but also of gas and coal, without accounting for an increase in the use of waste fuels such as waste wood. Net energy savings reflect the increase in waste fuel consumption. Although most of the early energy savings will be oil, over time the fuel savings from cogeneration will be mainly coal and nuclear, which are the fuels saved from not building as much new electric power baseload capacity. However, for 1985, we have assumed that additional electric utility capacity will not be delayed or cancelled as a result of cogeneration development, and that the utilities will produce less power from their high-cost plants, which would be mainly oil-fired. More specifically:

- Because of technical and economic constraints, without government action, process steam cogeneration development in the six industries analyzed will achieve only 35% of its 1985 maximum technical potential (or about 1,250 trillion Btu out of a possible 3,500 trillion Btu of process steam), providing an additional 6,000 MW of electric power capacity. This reverses the historical (decreasing) trend of cogeneration that existed prior to the 1970s.

- The technically suitable but economically infeasible cogeneration represents an opportunity for government action in the near term. Government action could generate an additional 2,000 to 6,000 MW from process steam cogeneration by 1985, resulting in as much as 300 trillion Btu of gross energy savings.

In addition, the technical constraints to cogeneration represent further opportunities for government action in the longer term (beyond 1985). However, the potential energy savings cannot be reliably calculated without further analysis. Moreover, accurate estimates of future potential would have to take into consideration the possibility of utility ownership and/or operation of the cogeneration facility.

FIGURE 3.1: PROCESS STEAM COGENERATION* FOR SIX INDUSTRIES (trillion Btu)

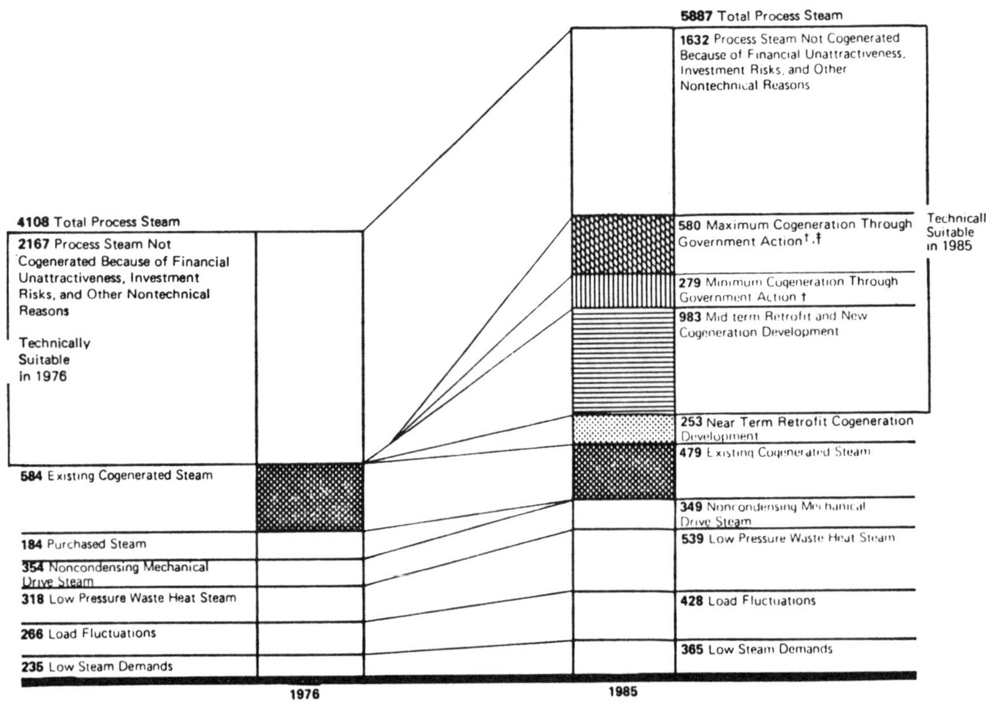

*Direct process heat topping cogeneration not shown.
†Government action begins in 1978.
‡In addition to amount shown under minimum government action.

SECTION 1
COGENERATION DEVELOPMENT WITHOUT GOVERNMENT ACTION

Since the Arab oil embargo and attendant price rises of 1973–1974, the economics of energy use have changed substantially enough to reverse the historical, declining trend of industrial cogeneration in the United States. These "new energy economics" will have the effect of more than doubling the current level of industrial cogeneration without any special government action. Specifically, it is estimated that, without government action, 1,700 trillion Btu (all numbers in this report are rounded) of process steam and almost 80 billion kWh of electric energy are likely to be cogenerated by 1985 in the six industries we examined (see Tables 3.5 and 3.6). The electrical output represents about 10,000 MW of cogeneration capacity, compared to a current level of somewhat more than 4,000 MW.

To arrive at these estimates of the cogeneration potential that will be realized by 1985, three markets for process steam were first defined:

> The near-term retrofit market represents the cogeneration potential at those manufacturing facilities which have existing process steam generation capacity. At these facilities, consideration has recently been given to the installation of cogeneration capacity. Although a portion of the potential development is technically and economically feasible, and no other barrier to development exists, it will not become operational until sometime after 1976.
>
> The mid-term retrofit market represents the cogeneration potential at those manufacturing facilities which have existing process steam generation capacity. However, cogeneration plants are not currently feasible at these facilities because of technical, economic, environmental, regulatory, or institutional reasons. Management will consider cogeneration capital investments in the 1977–1985 period. The cogeneration potential that is feasible for capital investment will be installed over this time span.
>
> The new development market represents the cogeneration potential at new facilities which will re-

TABLE 3.5: PROCESS STEAM COGENERATED BY 1985 IN SIX INDUSTRIES WITHOUT GOVERNMENT ACTION (trillion Btu)

Industry	Process Steam Cogeneration 1976	Phased Out During 1977-1985	Near Term Retrofit*	Mid-term Retrofit*	New Development	Process Steam Cogeneration 1985
Food	27	(5)	0	7	20	49
Textiles	15	(3)	0	0	1	13
Pulp & Paper	236	(42)	179	153	365	891
Chemical	190	(34)	32	56	164	408
Petroleum Refining	15	(3)	10	31	54	107
Steel	101	(18)	32	15	117	247
Total	**584**	**(105)**	**253**	**262**	**721**	**1,715**

*Retrofit, as used in this report, describes those plants with existing low pressure process steam generation capability in which electrical power is purchased. These plants can be replaced by a new cogeneration facility made up of a new high pressure steam generator and a turbine/generator plant.

quire steam generation capacity from 1977 to 1985. The portion of the potential cogeneration development that is technically and economically feasible will be installed over this period.

TABLE 3.6: ESTIMATED OUTPUT AND CAPACITY OF COGENERATION PLANTS IN SIX INDUSTRIES

	1985 Output		Capacity	
Industries	**Process Steam** (trillion Btu)	**Electric Energy** (billion kWh)	**1985 Total Cogeneration Capacity** (MW)	**1976 Total Cogeneration Capacity** (MW)
Food	49	2.1	343	189
Textiles	13	0.6	98	114
Pulp & Paper	891	38.3	4,861	1,286
Chemical	408	21.1	2,677	1,451
Petroleum Refining	107	6.0	763	118
Steel	247	11.2	1,423	887
Total	**1,715**	**79.3**	**10,165**	**4,045**

Having defined the three areas of market potential, the amount of total steam demand by each industry was then projected. The uses of this steam were then examined, excluding from cogeneration development steam already used for power production and steam found technically unsuitable for supply by a cogeneration plant; of the remaining steam demand, the amount associated with unattractive capital investment conditions (using a return on investment analysis) was excluded.

Six industries were analyzed: chemical, petroleum refining, paper and pulp, steel, food processing, and textile. For these industries major findings, by 1985, are given below:

> Of the approximately 6,400 trillion Btu of steam required annually, about 500 trillion Btu would be consumed by power plants (including cogeneration plants) already existing in 1976.
>
> Of the roughly 5,900 trillion Btu remaining, more than 40% is technically unsuitable for supply by a cogeneration plant due to operational and other conditions or is already supplied by existing cogeneration plants.
>
> Of the roughly 3,500 trillion Btu of technically suitable steam in demand, almost 70% corresponds to

capital investment conditions that are unattractive to the industries examined.

Approximately 1,000 trillion Btu of the 1985 annual steam consumption were found to be both technically suitable and financially attractive for investment by industry in new cogeneration plants over the 1977–1985 period, both as additions to existing manufacturing facilities, as well as integral parts of new manufacturing facilities. This amount is in addition to the 250 trillion Btu of process steam suitable for investment in 1976. In total, the most economically attractive cogeneration configurations for this supply of steam would provide approximately an additional 6,000 MW of capacity by 1985.

Specific findings for cogeneration development in each of the six industries are summarized in Table 3.7. As the exhibit shows, the industry with the greatest increase in and amount of cogeneration capacity by 1985 is expected to be pulp and paper (about 48% of the total in 1985 compared to 32% in 1976). In contrast, the textile industry is anticipated to have a net decline in cogeneration capacity over the period and contribute the least amount to the 1985 total (about 1% of the total in 1985 compared to 3% in 1976).

TABLE 3.7: ESTIMATED COGENERATION BY INDUSTRY IN 1976 AND 1985 WITHOUT GOVERNMENT ACTION

	Process Steam Cogenerated (trillion Btu)		Electric Energy Cogenerated (billion kWh)		Cogeneration Capacity (MW)	
Industry	1976	1985	1976	1985	1976	1985
Food	27	49	1.16	2.10	189	343
Textiles	15	13	0.69	0.60	114	98
Pulp & Paper	236	891	10.14	38.33	1,286	4,861
Chemicals	190	408	11.44	21.08	1,451	2,677
Petroleum Refining	15	107	0.94	6.02	118	763
Steel	101	247	7.00	11.19	887	1,423
Total	**584**	**1,715**	**31.37**	**79.32**	**4,045**	**10,165**

Specific findings for cogeneration development by plant type are summarized in Table 3.8. As shown in the table, waste fuel-fired cogeneration capacity is expected to constitute the greatest portion of 1985 cogeneration capacity (about 47%) because of the increased emphasis on process residual fuel consumption, especially in the pulp and paper industry. Because of the mandatory coal utilization program, cogeneration plants fired by coal will make up the largest portion of plants using purchased fuels.

TABLE 3.8: ESTIMATED COGENERATION BY PLANT TYPE IN 1976 AND 1985 WITHOUT GOVERNMENT ACTION

Cogeneration Plant Type	Process Steam Cogenerated (trillion Btu)		Electric Energy Cogenerated (billion kWh)		Cogeneration Capacity (MW)	
	1976	1985	1976	1985	1976	1985
Coal-fired steam turbine	97	502	4.16	21.58	550	2,791
Residual-fired steam turbine	204	231	8.81	9.95	1,157	1,294
Distillate-fired gas turbine	20	23	3.51	4.02	446	524
Natural gas turbine	65	52	4.34	3.51	554	449
Waste fuel steam turbine	182	851	9.90	37.85	1,254	4,800
Heat recovery steam turbine	16	56	0.65	2.41	84	302
Total	**584**	**1,715**	**31.37**	**79.32**	**4,045**	**10,165**

Figure 3.2 illustrates the steam supply and usage factors considered. This figure should be referred to as necessary to understand discussions in this section.

STEAM UNSUITABLE FOR CAPITAL INVESTMENT CONSIDERATION

Analysis of a number of factors indicates that 60 to 70% of the total process steam consumption in the six industries examined is unsuitable for capital investment consideration. Specifically, of the approximately 6,400 trillion Btu of steam consumption estimated in 1985 for the six industries examined, about 4,000 trillion Btu will be unsuitable for a cogeneration plant investment consideration, due to technical and nontechnical factors. In brief, these factors fall into three categories: steam already consumed in power plants (cogeneration and other); technical constraints that limit the amount of steam supplied by a (baseload) cogeneration plant; and nontechnical factors involving an-

FIGURE 3.2: TECHNICAL AND FINANCIAL FACTORS INVOLVED IN THE GENERATION AND USE OF STEAM IN INDUSTRY

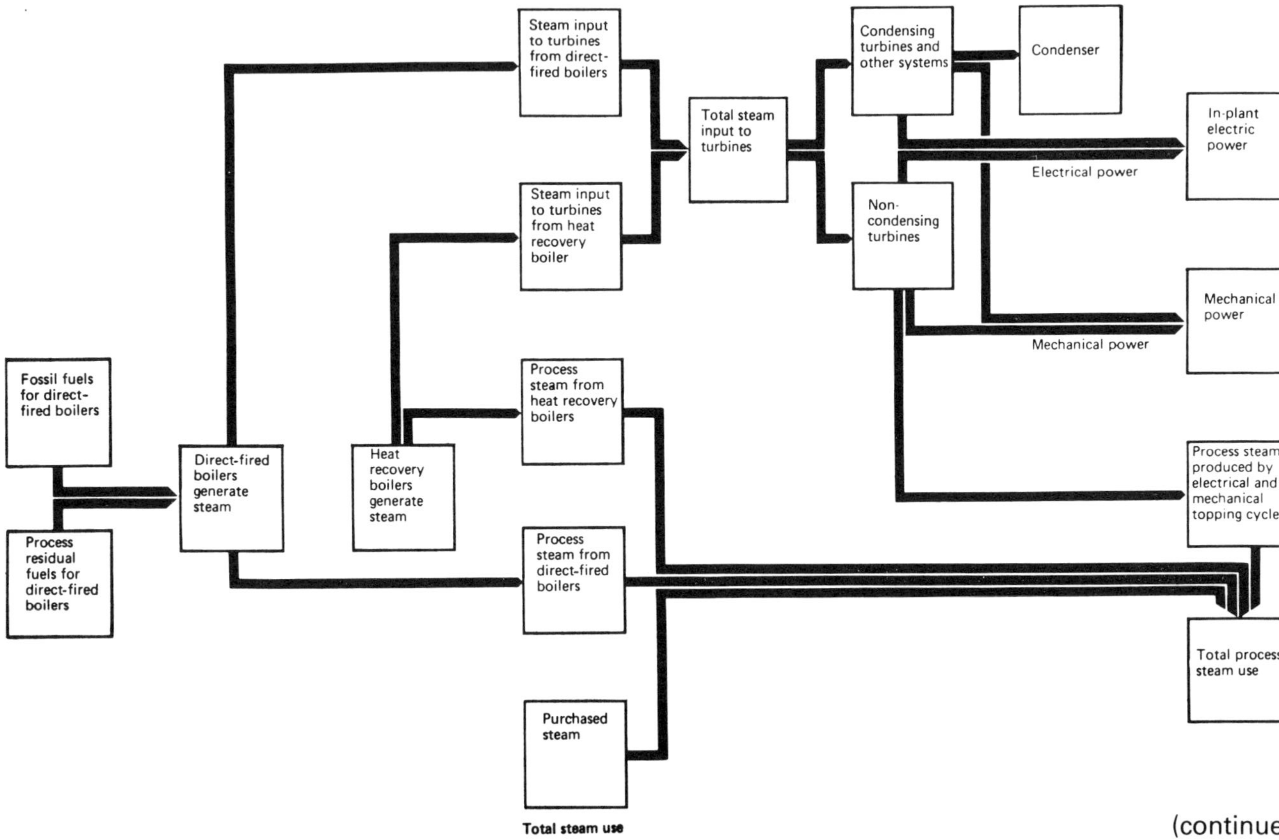

(continued)

FIGURE 3.2: (continued)

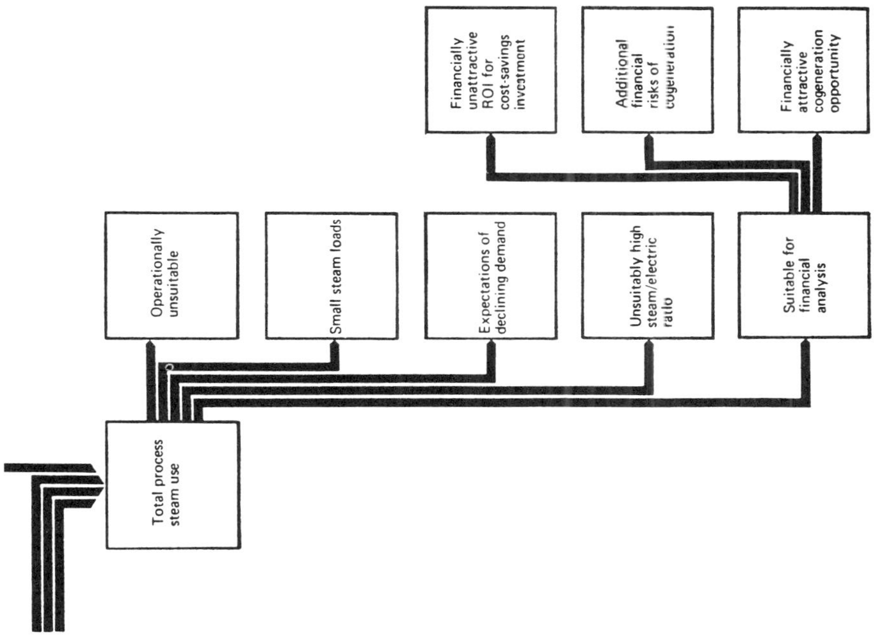

ticipated declining demand and steam/electric ratios. The impact of each of the factors on cogeneration development is summarized in Table 3.9.

TABLE 3.9: COGENERATION POTENTIAL TECHNICALLY AND ECONOMICALLY SUITABLE FOR DEVELOPMENT
(thermal value of steam in trillion Btu)

	1976	1985		
	Near Term Retrofit Market	Mid-Term Retrofit Market	New Development Market	Total
Total steam demand	**4734**	**3901**	**2505**	**6406**
Steam used for power production	(626)	(508)	(11)	(519)
Process steam demand	**4108**	**3393**	**2494**	**5887**
Technical constraints:				
Existing cogenerated steam	(584)	(479)	–	(479)
Planned cogenerated steam	–	(253)	–	(253)
Noncondensing mechanical drive steam	(354)	(291)	(58)	(349)
Purchased steam	(184)	–	–	–
Load fluctuations	(266)	(201)	(227)	(428)
Low steam demands	(235)	(179)	(186)	(365)
Low pressure waste heat steam	(318)	(362)	(177)	(539)
Process steam technically unsuitable for cogeneration	**(1941)**	**(1765)**	**(648)**	**(2413)**
Technically suitable process steam demand	**2167**	**1628**	**1846**	**3474**
Nontechnical constraints:				
Declining demand expectations	(434)	(397)	(63)	(460)
Inappropriate steam/electric ratio	(429)	(324)	(331)	(655)
Process steam suitable for capital investment consideration	**1304**	**907**	**1452**	**2359**
Financial constraints:				
Unacceptable ROI	(1044)	(625)	(654)	(1279)
Unacceptable risk	(7)	(20)	(77)	(97)
Process steam suitable for capital investment	**253**	**262**	**721**	**983**

Steam Used for Power Production

To estimate process steam demand in the six industries studied, the total steam demand was adjusted by the amount of steam consumed in power production. The total steam demand is supplied by direct-fired boilers (fossil and process residual fuels), waste heat recovery boilers, and purchases. From examination of these sources, a total steam demand in 1976 of 4,700 trillion Btu in all six industries was estimated. For each industry, the elements making up total 1976 steam supply are presented in Table 3.10.

TABLE 3.10: TOTAL STEAM USE BY INDUSTRY IN 1976 (trillion Btu)

Industry	Generated by Direct-Fired Boilers	Generated by Waste Heat Recovery Boilers	Purchased Steam	Total Steam Use
Food	344	8	0	**352**
Textiles	139	0	0	**139**
Pulp & Paper	1,361	0	18	**1,379**
Chemical	1,416	200	123	**1,739**
Petroleum Refining	424	155	43	**622**
Steel	463	40	0	**503**
All six industries	4,147	403	184	**4,734**

To project total steam demand by 1985, industry growth, existing manufacturing capacity phaseouts, and energy conservation improvements were estimated. Growth and conservation vary by industry and affect the amount of projected 1985 steam demand, which was estimated at 6,400 trillion Btu. Estimates for each industry are summarized in Table 3.11.

TABLE 3.11: FORECASTING ASSUMPTIONS FOR TOTAL 1985 STEAM USE FOR SIX INDUSTRIES

Industries	**Annual Growth Rate 1976-1985** (percent)	**Percentage of 1976 Energy Use Conserved by 1985**	**Total Steam Use in 1985** (trillion Btu)
Food	4.10	15	455
Textiles	3.20	27	160
Pulp & Paper	3.60	26	1,838
Chemical	5.25	19	2,466
Petroleum Refining	2.50	16	778
Steel	2.85	12	709
Total			**6,406**

Of the 6,400 trillion Btu total, over 500 trillion Btu will be consumed in the production of electrical and mechanical power by power plants existing in 1976. Currently, existing cogeneration electric power plants consume over 100 trillion Btu of steam in the production of electric power. Accounting for some likely phaseouts of existing equipment, this figure will drop to well under 100 trillion Btu by 1985. Second, steam is also used in noncondensing mechanical drive turbines (e.g., pumps and compressors). Third, some of the steam energy is consumed in condensing turbines in electrical as well as mechanical systems. Over 400 trillion Btu of steam is expected to be consumed in the last two categories.

The amount of steam energy consumed in power production varies by industry. For the petroleum refining industry, power production represents 6% of total steam use, whereas for the steel industry, this power production represents about 28% of total steam use. In the remaining four industries examined, power production represents from 10 to 13% of total steam use.

Technical Constraints On Cogeneration Development

Analysis of the 5,900 trillion Btu of process steam available for cogeneration development in 1985 shows that about 2,400 trillion Btu would not be suitable for supply by a cogeneration plant because of the following technical factors: steam is already supplied by electrical cogeneration plants (noncondensing systems); steam is already supplied by mechanical drive (noncondensing systems); steam is purchased; load fluctuations are too great; steam loads are too small; or waste heat boiler steam pressure is too low.

Existing Electrical Cogeneration Plants: In 1976, the six industries examined produced about 600 trillion Btu of process steam from cogeneration plants. Assuming a 2% per year phaseout of this capacity, by 1985 this steam energy should drop by about 100 trillion Btu (as shown in Table 3.9), from 584 to 479 trillion Btu. Further, about 250 trillion Btu of process steam will be produced in 1985 by cogeneration plant investments currently attractive but not yet implemented (i.e., that portion of steam both technically suitable and financially attractive for immediate cogeneration investments).

Most of the existing development and attractive potential cogeneration is in the pulp and paper industry, which cogenerated about 240 trillion Btu of process steam in 1976. This industry is expected to develop new cogeneration capacity that will produce an additional 180 trillion Btu of process steam. In contrast, it is estimated that the petroleum refining and textile industries have a current cogeneration capacity of only 15 trillion Btu each of process steam. For the refining industry, about 10 trillion Btu of process steam is technically suitable and financially attractive for an immediate investment; no cogeneration investment is likely for existing plants in the textile industry.

Noncondensing Mechanical Drive Systems: Noncondensing mechanical drive systems produced about 350 trillion Btu of process steam in 1976 and are estimated to produce a similar amount in 1985. This represents a direct reduction in the potential for the cogeneration of electric power and process steam. Some industries rely on noncondensing mechanical drives more than others. The steel industry, in fact, accounts for about 28% (106 trillion Btu) of process steam generated by these systems and is expected to increase its use of mechanical drive equipment by 15% over the 1977–1985 period. The other five industries studied, however, are expected to decrease their use by 18% over the same period, based on our industry interviews.

In general, the technical and economic competitiveness of electric equipment is expected to decrease industry's dependence on mechanical drive systems. Mechanical drive systems, however, do have some advantages over electric motors. They are more easily adapted to adverse operating environments, such as high temperatures and high particulate concentrations. Also, they are more responsive to variable speed requirements; and they are more easily adapted to small spaces when high power—over 1,000 horsepower—is required.

Purchased Steam Constraints: Currently, some industries purchase some of their steam requirements (the chemical industry, for example, purchases about 120 trillion Btu). These purchases are under contracts of varying time periods, and generally cannot be arbitrarily terminated when a cogeneration plant is to be installed. Therefore, purchased steam reduces the immediate potential for cogeneration. By 1985, however, it was assumed that this amount would be zero for purposes of this analysis (see Table 3.9).

Steam Load Fluctuations: Fluctuations in steam (and electrical) demand are caused by certain manufacturing operations. Cogeneration systems, however, require a fairly steady and balanced demand for steam and electricity. For the six industries examined, the overall reduction in cogeneration steam potential because of load fluctuations was estimated to be 270 trillion Btu in 1976, and about 420 trillion Btu in 1985. In 1985 projections, it was assumed that no significant changes in manufacturing processes and related operations would mitigate this factor. At new manufacturing facilities, however, this technical constraint might be reduced if technological innovations related to load smoothing were developed.

There are several ways in which load fluctuations are now handled by industry. First, extraction condensing turbines are used to allow a diversion of reduced process steam demand to the condensors where the energy is exhausted through the cooling system to the environment (similar to an electric utility condensing turbine plant). Also, standby steam boilers are available to provide for additional process steam that may be periodically needed. In addition, most manufacturing concerns which cogenerate power purchase supplemental power from a utility when their extraction condensing turbines are experiencing a high steam demand with a reduction in internally generated electric power, or when noncondensing turbines are incapable of meeting increased electric demand.

The reduction of cogeneration potential from load fluctuation varies by industry. The steel and petroleum refining industries incur relatively minor fluctuations in steam and electric demand. The chemical and paper industries have a somewhat less stable demand, and the textile and food industries experience significant daily changes and seasonal steam demands.

Low Steam Demand: Low steam demand in some manufacturing plants cannot be efficiently served by cogeneration plants. For the noncondensing steam turbine cogeneration technologies addressed in this study, the minimum steam flow, using standardized system components without significantly reducing the system efficiency, is about 50,000 to 100,000 pounds per hour (less for a gas turbine with a waste heat recovery boiler than for a back-pressure turbine). In this study, it was assumed that a cogeneration plant using standardized components would not be technically efficient and viable below 50,000 pounds per hour steam demand at any single manufacturing plant. For the six industries examined, this constraint was estimated to approximate 240 trillion Btu currently (1976), and about 360 trillion Btu of processed steam in 1985. Although steam loads under 50,000 pounds per hour are not strictly infeasible, inadequate operator expertise at small plants and fewer economies of scale make such small plants unattractive from an economic standpoint.

It has been assumed that the fraction of each industry's steam demand attributable to small steam loads remains the same through 1985. These estimates may be high to the extent that new manufacturing facilities will be larger on the average than existing manufacturing facilities, and hence will require greater amounts of process steam. This would particularly affect the textile and food processing industries, where most of the small steam loads occur, accounting for about 25% of the total process steam demand in these industries. In contrast, in the steel and petroleum refining industries, all facilities were assumed to be sufficiently large to cause no reduction because of this technical constraint.

Low-Pressure Waste Heat Boiler Steam: Low-pressure waste heat recovery boilers are estimated to produce about 540 trillion Btu of process steam in 1985, approximately a 70% increase over 1976. This represents a direct reduction in the potential cogenerated steam, since to supply this steam from a cogeneration plant is to lose the full value of the waste heat captured.

The petroleum refining industry is estimated to generate the greatest amount of steam from waste heat: 28% of all steam is generated by waste heat recovery boilers. At the other extreme, no waste heat-generated steam was estimated for the pulp and paper industry because of the nature of the processes.

Between now and 1985, two trends are at work affecting the amount of waste heat-generated steam. On one hand, at existing and new manufacturing facilities, energy economics now makes more waste heat recovery boilers financially attractive. As has been already mentioned, some of the waste heat steam generated will be at a sufficient pressure to generate electricity in a condensing or noncondensing turbine. On the other hand, energy-saving technologies will be implemented at new manufacturing facilities to save waste heat losses, with the potential effect of reducing both the input process steam needed and "bottoming" cogeneration potential.

Nontechnical Constraints on Cogeneration

Beyond the technical constraints to cogeneration, there are two other factors at work that remove cogeneration plant investments from any consideration by industrial managers. These are expectations of declining steam demand, and high steam-to-electric demand conditions that require a decision to export power to the utility system if a cogeneration investment is to be made. These constraints further reduce the cogeneration potential by about 1,100 trillion Btu in 1985.

Expectations of Declining Demand: Even when steam loads are currently technically suitable for cogeneration, management may not consider an investment in such facilities because of expected declining demand for process steam. Plant shutdowns and steam conservation programs decrease potential cogeneration development by approximately 430 trillion Btu in 1976; in 1985 markets, the decrease is expected to be 460 trillion Btu. Since the planned life of a cogeneration plant is 20 to 30 years, such plants will not be built for any manufacturing steam demand that would be scheduled for phaseout within a shorter (i.e., 10-year) time frame. Such an investment would obviously have poor economics.

Further, major steps are expected to be taken by industry to improve or replace some energy-consuming processes, waste heat losses and steam over the next 5 to 10 years. For instance, expected improvements in steam trap performance are likely to reduce the total steam required.

An assessment of the 1976 expectations of declining steam demand is incorporated into the total steam figures projected for 1985. This assumption varies significantly by industry, reflecting particular energy conservation targets in each industry.

Steam/Electric Ratio Investment Considerations: For some manufacturing facilities, the ratio of steam to electric power demand is sufficiently high that no configuration can produce a low enough electric output per pound of steam for internal power requirements. The highest steam/electric ratio looked at in this study uses a back-pressure turbine with a boiler pressure of 600 pounds per square inch gauge (psig). High steam/electric ratios have been accommodated in cases where the electric power demand is large and only a portion of the high steam demand can justify an economically-sized cogeneration plant. However, where the electric power requirement is small and the steam requirements large, the industrial manager must consider either the installation of an unusually small (and less economic) cogeneration plant, or sizing the cogeneration plant to match the manufacturing facilities' steam load, and selling the excess power to the utility system.

To assess what price the industrial cogeneration plant owner would have to receive for excess power in order to obtain the same financial return from replacement of his own purchased power, it was determined that the utility's price would have to be quite close (within 10 to 20%) to the price at which the company would purchase power from the electric utility. On the basis of electric utility and industrial company interviews, it is believed that such a price is probably not achievable without unusual tax or other incentives. The conclusion was reached that any cogeneration plant investment for internal power only would yield a higher return than the incremental investment to supply excess power. In the absence of a more comprehensive utility company investment analysis, and reflecting the strong reluctance by industrial managers interviewed for selling excess power under regulation, it was determined that industrial companies will not make cogeneration investments in these high steam electric demand cases.

Specifically, the estimate is that about 430 trillion Btu, or 10% of the process steam demand in the six industries analyzed, is unsuitable for cogeneration without sale of power. For 1985, it is estimated that about 660 trillion Btu or 11% of the process steam demand will fall into this category. The chemical and petroleum refining industries have the greatest fraction of their steam constrained by this factor. They therefore have the greatest potential for further cogeneration development as a result of any incentives to stimulate power sales to the utility system.

CAPITAL INVESTMENT CONSTRAINTS TO COGENERATION DEVELOPMENT

The most sensitive part of the estimation of 1985 cogeneration development without government action was the return on investment (ROI) analysis of the six industries examined. Industry discussions focused on criteria for investing in cogeneration plants, given a technically feasible opportunity. In total, these discussions covered companies representing over half of the total cogeneration potential in the six industries.

The impact of the return on investment and risk characteristics of cogeneration investments is summarized in Table 3.12. Without government action, it is estimated that less than 1,000 trillion Btu of process steam will be acceptable for cogeneration investments over the 1977–1985 period. This amount is in addition to the approximately 250 trillion Btu of the 1976 process steam demand currently suitable for capital investment. This total constitutes almost half of the steam suitable for investment consideration in 1985. There are two reasons why so much potential cogeneration is financially unattractive:

> Cogeneration investments often do not meet the normal return on investment criteria of the industries.
>
> In addition, uncertainties and risks associated with cogeneration plant investments—such as uncertainty over fuel price and availability, environmental standards, and government regulation—further limit industry's willingness to invest in cogeneration projects.

In arriving at the above findings, the return on investment for different configurations of cogeneration plants with different electricity and fuel prices (to allow a regional analysis) was determined. These ROIs were matched to the corporate "hurdle rates" (minimum acceptable ROIs) for the industries examined. Although other financial criteria are used in addition to the hurdle rate (e.g., pay-back), nearly all companies interviewed in this study used ROI as the single most important criterion.

TABLE 3.12: COGENERATION POTENTIAL SUITABLE FOR CAPITAL INVESTMENT (thermal value of steam in trillion Btu)

	1976	1985		
	Near Term Retrofit	Mid-Term Retrofit Market	New Development Market	Total
Process steam suitable for capital investment consideration	**1304**	**907**	**1452**	**2359**
Financial constraints:				
Unacceptable ROI	(1044)	(625)	(654)	(1279)
Unacceptable risk	(7)	(20)	(77)	(97)
Process steam suitable for capital investment	**253**	**262**	**721**	**983**
Total process steam produced by cogeneration plants in the six industries	**584**	**994**	**721**	**1715**

Return on Investment

The limited returns on cogeneration investment, compared to normal investment hurdle rates (Figure 3.3), reduce the potential for cogeneration development by over 1,000 trillion Btu in 1976 and about 1,300 Btu by 1985, due to a combination of high capital costs and low cost savings of cogeneration plants in certain industries or regions.

FIGURE 3.3: COGENERATION POTENTIAL THAT WILL DEVELOP UNDER NORMAL ROI HURDLE RATES*

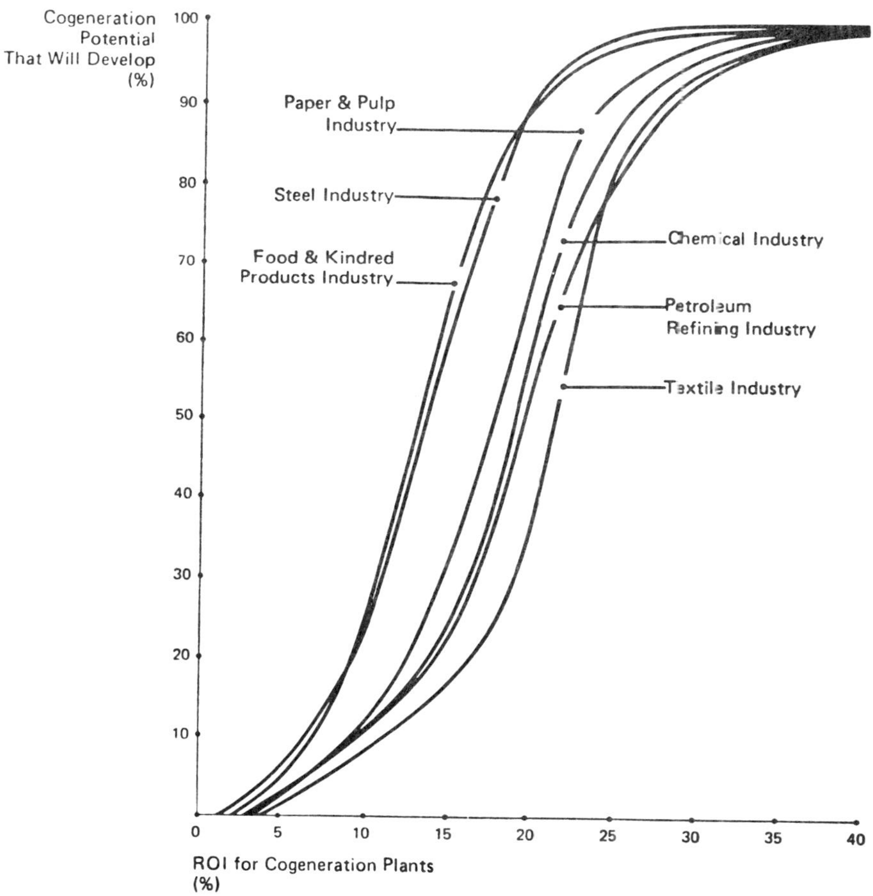

*Industry interview results showed that some companies do make energy conservation investments at low or negligible ROIs for policy, modernization, and other reasons.

To determine the financial attractiveness of cogeneration investments, the after-tax ROIs for cogeneration projects were constructed in a manner comparable to that used by managers in each of the six industries as described below.

Table 3.13 presents the various options considered in this study to be available to management decision makers. In a retrofit application, i.e., for eventual modification of existing steam generation facilities, the after-tax ROIs were determined as the incremental attractiveness of a new cogeneration plant over the continued operation of existing steam generating capacity and purchase of electric energy. In the new development market, i.e., for meeting new requirements of process steam between 1976 and 1985, the after-tax ROIs were determined as the incremental attractiveness of a new cogeneration plant over the installation of a new boiler and the purchase of the required energy. In all instances, the cogeneration plants and boilers were sized equally and matched to the expected baseload steam flow demand. Because management does not anticipate generating excess electric power without government action as discussed in the previous section, no cogeneration plant was allowed to exceed the baseload electric power demand.

TABLE 3.13: COGENERATION OPTIONS CONSIDERED FOR INCREMENTAL ROI ANALYSES

Markets	Options
Near term and mid-term retrofit markets	An oil-fired cogeneration plant versus an existing oil-fired boiler.
	A coal-fired cogeneration plant versus an existing coal-fired boiler.
	A coal-fired cogeneration plant versus an existing oil-fired boiler in the absence of the coal conversion requirement (voluntary conversion).
	A coal-fired cogeneration plant versus an existing oil-fired boiler in the case of the coal conversion requirement (mandatory conversion).
	A distillate oil-fired gas turbine with a waste heat recovery boiler versus an existing oil-fired boiler.
New development markets	An oil-fired cogeneration plant versus an oil-fired boiler.
	A coal-fired cogeneration plant versus a coal-fired boiler.
	A distillate oil-fired gas turbine with a waste heat recovery boiler versus an oil-fired boiler.

NOTE: The incremental ROI is the result of an analysis that compares the differences between the benefits and costs of the two options being considered for capital investments.

Table 3.14 presents the distribution of the cogeneration options assumed for each industry in 1976 and 1985. As an example, in the food processing industry, companies which account for 15% of the total cogeneration potential considered for capital investments in 1976 and which presently use coal for process steam generation would consider a coal-fired cogeneration plant. The values assigned to the distributions of these cogeneration options for each industry in each market are based on historical fuel use information, the regional distribution of manufacturing plants, and the expectations and plans of representatives within each industry. This seems to be a reasonable representation of the cogeneration fuel types under consideration by industrial managers.

TABLE 3.14: DISTRIBUTION OF COGENERATION FUEL TYPES CONSIDERED FOR INVESTMENT BY EACH INDUSTRY

	Percent Distribution by Industry					
Market Application and Configuration	**Food**	**Textiles**	**Pulp & Paper**	**Chemicals**	**Petroleum Refining**	**Steel**
Near Term Market						
Coal-fired steam turbine						
—no conversion	15	15	10	15	1	10
—voluntary conversion	15	15	30	31	10	10
—mandatory conversion	0	0	0	0	0	0
Oil-fired steam turbine	50	50	15	35	54	8
Oil-fired gas turbine with waste heat recovery boiler	20	20	5	10	10	7
Process residual fuel steam turbine	0	0	40	4	15	60
Waste heat recovery steam turbine	0	0	0	5	10	5
Total	**100**	**100**	**100**	**100**	**100**	**100**
Mid-Term Market						
Coal-fired steam turbine						
—no conversion	15	15	10	15	1	10
—voluntary conversion	0	5	0	0	0	3
—mandatory conversion	15	30	20	35	29	10
Oil-fired steam turbine	50	30	8	30	25	5
Oil-fired gas turbine with waste heat recovery boiler	20	20	2	10	5	7
Process residual fuel steam turbine	0	0	60	5	25	60
Waste heat recovery steam turbine	0	0	0	5	15	5
Total	**100**	**100**	**100**	**100**	**100**	**100**
New Development Market						
Coal-fired steam turbine (mandatory coal use)	40	55	20	57	40	21
Oil-fired steam turbine	40	20	8	20	10	2
Oil-fired gas turbine with waste heat recovery boiler	20	25	2	10	5	7
Process residual fuel steam turbine	0	0	70	8	30	65
Waste heat recovery steam turbine	0	0	10	5	15	5
Total	**100**	**100**	**100**	**100**	**100**	**100**

Because of the importance of coal conversion to cogeneration development, the impact of DOE's recommendation to the U.S. Congress that coal be used for steam generation producing over 100 million Btu per hour of thermal energy was carefully considered. A differentiation was made between the voluntary and mandatory consideration by industry decision makers to use coal instead of other fossil fuels. Although some cogeneration plants now under development will be mandated to use coal, it was assumed that the estimate of 1976 potential would not be affected by the mandatory conversion program.

From the options available to industry decision makers the following situations can be conducive to cogeneration capital investments:

> In the retrofit market, if industry were required to use coal as a boiler fuel, existing oil or gas boilers could be replaced by either a new coal-fired boiler or a coal-fired cogeneration plant. The ROI of the latter may be improved because of the relatively low additional capital cost required for the cogeneration plant to achieve significant energy cost savings.
>
> In the retrofit market, the voluntary conversion from oil to coal generally does not yield as high an ROI as does the mandatory conversion because the capital costs of coal-fired cogeneration are incremental to those of an oil-fired boiler. A ready source of coal, particularly clean coal, would, however, improve the willingness of operators to build new steam cogeneration facilities because coal is generally less expensive per million Btu than oil.
>
> In the new development market, plans to add manufacturing capacity must include plans for new steam generation. In this case, the incremental capital required for cogeneration over that for a new boiler may be sufficiently small so that the ROI is attractive.

Despite these situations in which cogeneration ROI can be attractive, the overall potential for development is limited. The following section discusses, for the six industries studied, the financial factors that limit cogeneration development: the high capital costs and the low cost savings of cogenerated steam and electricity compared with the alternative of process steam generation and the purchase of electricity.

High Capital Costs: The capital costs of cogeneration plants are related to the design requirements, that is, the type of configuration and its size. In general, industrial cogeneration plants have higher capital costs than industrial boilers because of the need for more expensive emission control, electrical, and boiler equipment. For example, a 30-MW coal-fired system with a straight back-pressure noncondensing steam turbine could cost more than $29 million, or about two-thrids more than a coal-fired process steam boiler with the same steam generating capacity.

The desired cogeneration plant type depends upon the ratio of electric power to process steam capacity required to replace, either in part or totally, the existing purchased electric power and process steam capacity. For steam turbine systems of the same fuel type with comparable process steam pressures and flows, but different electric power capacity, the investment requirement is greater for the cogeneration plant with the greater capacity. These increased capital costs are caused by the higher boiler pressures and the larger turbine/generator required to produce a greater amount of electric power.

For cogeneration plants with comparable steam and electric capacities, the investment required for a coal-fired plant is considerably greater than that for an oil-fired plant. Although some of the capital cost differences are attributed to the additional environmental requirements for the coal-fired steam generation (e.g., land area and emission controls), the boilers themselves are the major contributors to the cost differences. For instance, a high-pressure coal-fired boiler of 1,500 psig used in some industrial cogeneration plants has a capital cost about 1.4 times that of a comparable oil-fired boiler.

The investment requirements are disproportionately high for small cogeneration plants, thereby producing low ROIs for small cogeneration plants as illustrated in Figure 3.4. Although this effect is attributed to the lack of economies of scale for electric devices, boiler equipment, and most system components, one of the major contributors is the steam turbine generator. For instance, a 5-MW noncondensing steam turbine generator with turbine inlet steam of 850 psig and 825°F has a cost per MW about 2.3 times greater than a 30-MW turbine.

Although the steam and electric requirements of cogeneration plants vary throughout each industry, analysis of these requirements for baseload operation and the characteristics of present and planned cogeneration plants indicates that the use of typical cogeneration parameters for each industry is reasonable. These values are presented in Table 3.15.

FIGURE 3.4: INCREMENTAL ROI OF COAL-FIRED COGENERATION PLANTS IN DIFFERENT MANUFACTURING APPLICATIONS

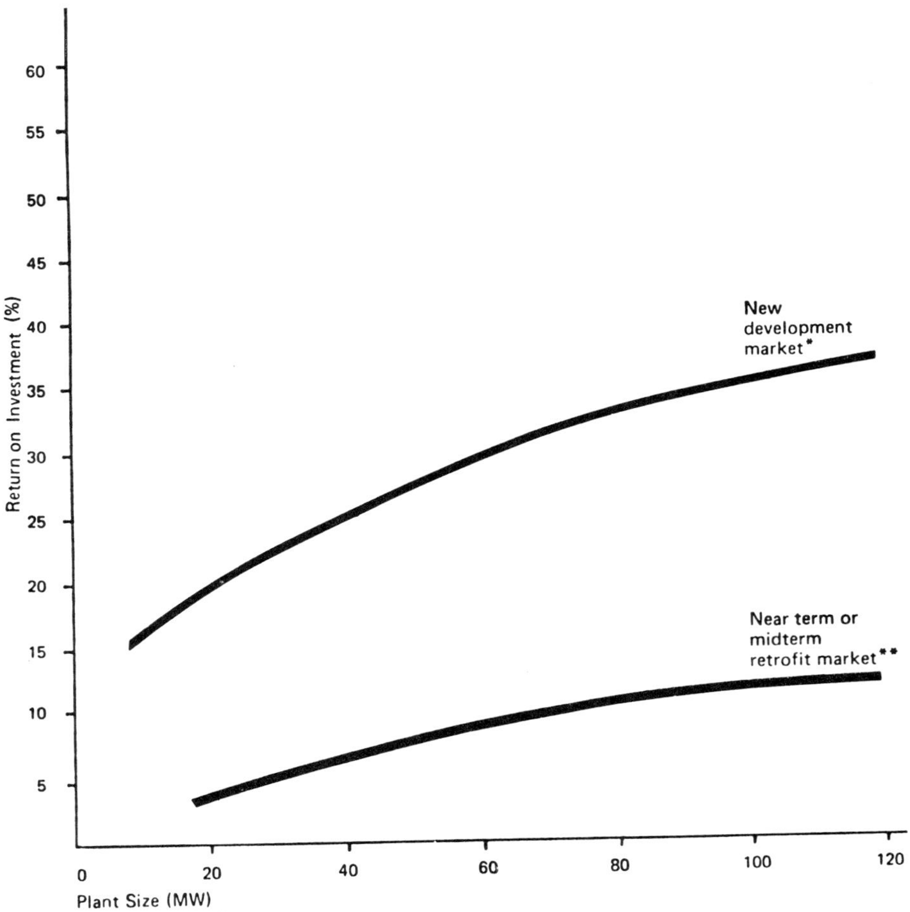

*The incremental ROI is that of a new coal-fired cogeneration plant compared to a new coal-fired process steam generating plant and the purchase of electric power.

**The incremental ROI is that of a new coal-fired cogeneration plant compared to the continued use of an existing coal-fired process steam generating plant and the purchase of electric power.

Inlet condition: 850 psig/825°F.
Exhaust condition: 150 psig.
Annual capacity factor: 0.90.
Fuel price: $1.40/MMBtu.
Electricity price: 2.7¢/kWh
Annual standby power cost: $50/kW for ½ of plant capacity.

TABLE 3.15: COGENERATION DEVELOPMENT PARAMETERS

Plant Size

Market	Food	Textiles	Pulp & Paper	Chemical	Petroleum Refining	Steel
Near term retrofit (% of total)						
5-20 MW	90	90	40	40	20	20
21-40 MW	10	10	20	20	20	20
41-80 MW	0	0	20	20	20	40
>80 MW	0	0	20	20	40	20
Mid-term retrofit (% of total)						
5-20 MW	90	90	30	30	10	20
21-40 MW	10	10	30	30	20	20
41-80 MW	0	0	20	20	30	40
>80 MW	0	0	20	20	40	20
New development (% of total)						
5-20 MW	85	80	30	20	10	10
21-40 MW	15	20	25	40	10	20
41-80 MW	0	0	25	20	30	40
> 80 MW	0	0	20	20	50	30

Physical Characteristics

Process steam pressure for major steam load (psig)	50	50	150	150	150	150
Turbine inlet steam conditions*						
5-40 MW	low	medium	medium	medium	medium	medium
> 40 MW	low	medium	high	high	high	high

*Turbine inlet steam conditions used in the analyses are:
Low—600 psig/750°F
Medium—850 psig/825°F
High—1450 psig/950°F.

Because the food and textile industries generally require small cogeneration plants, the disproportionately high capital cost incurred would result in low ROIs for plants in these industries. All industries are affected by the high capital costs associated with coal-fired cogeneration plants through generally low ROIs. Although the industries where high electric-to-steam ratios are required will incur greater capital costs, the ROIs may be higher due to a great displacement in purchased electricity, producing larger costs benefits.

Low Cost Savings: The cost savings offered by cogeneration plants vary according to the capacity utilization of the plant, the fuel costs compared to electric energy costs, and the charges for utility standby services. Each of these factors is addressed below.

Low capacity utilization can limit the financial attractiveness of cogeneration. The annual utilization of cogeneration capacity depends fundamentally on the process steam and electric power demand of the manufacturing facility throughout the year. Variations in manufacturing activity—daily, weekly, or seasonally—are reflected in variations in energy demand. In turn, the energy demand influences the amount of cost benefits that can be derived through cogeneration. As illustrated in Figure 3.5, a greater annual utilization results in a higher ROI.

For each industry studied, a single annual utilization rate was assigned to meet the most constant and basic requirements (baseload operation). The utilization rates presented in Table 3.16 were assigned primarily on the basis of industry interviews. The textile and food industries are characterized by low annual utilizations because of their generally nonconstant manufacturing activity. For instance, integrated textile mills and finishing plants generally operate five to six days a week with two shifts of manufacturing activity. The food industry, which is very diverse in character, has some plants that operate 24 hours a day for only about half the year, depending on agricultural activity; others operate five to six days a week with one to three shifts of operation for the whole year. In contrast, integrated manufacturing plants in the steel, petroleum refining, and paper industries as well as the majority of cogeneration candidates in the chemical industry generally operate up to 24 hours a day, 365 days a year, which offers the opportunity for high annual utilization.

TABLE 3.16: OPERATIONAL FACTORS INFLUENCING THE INCREMENTAL SAVINGS OF COGENERATION PLANTS

Operational Factor	Food	Textiles	Pulp & Paper	Chemical	Petroleum Refining	Steel
Purchased electric price (¢/kWh)*	2.6	2.4	2.5	2.5	2.6	2.5
Fuel prices ($/million Btu)*						
Coal	1.25	1.17	1.23	1.43	1.63	1.11
Residual oil	2.20	no discernible differences in residual oil prices				
Distillate oil	2.40	no discernible differences in distillate oil prices				
Annual utilization (%)	70	70	90	90	90	90

*Based on geographic distribution of industries and weighted regional analyses.

FIGURE 3.5: INCREMENTAL ROI FOR COAL-FIRED COGENERATION PLANTS WITH VARYING ANNUAL UTILIZATION AT EXISTING MANUFACTURING FACILITIES*

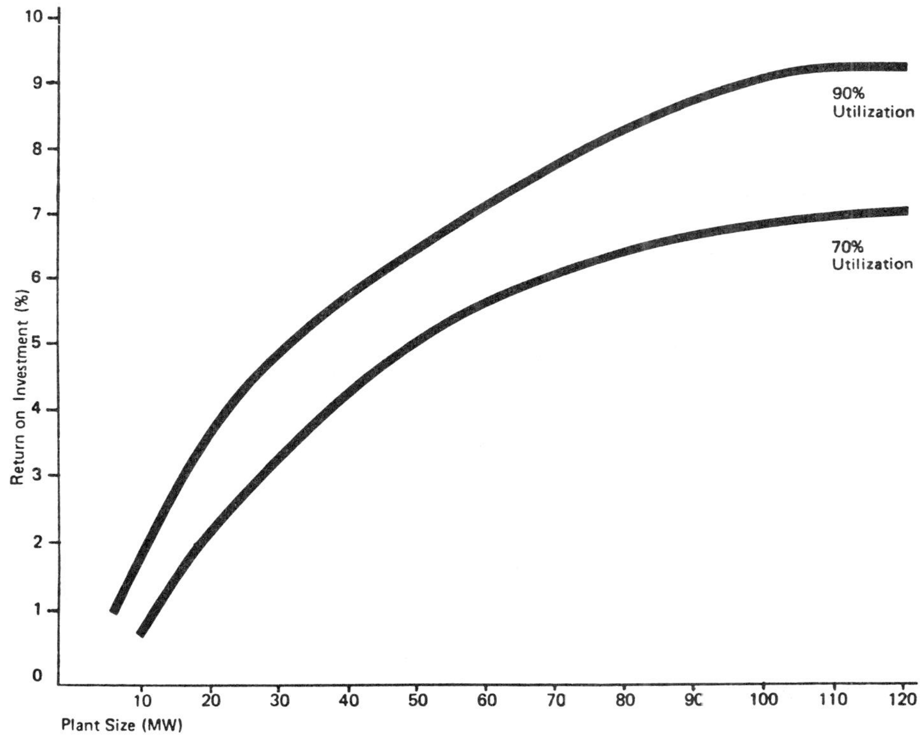

*The incremental ROI is that of a new coal-fired cogeneration plant compared to a new coal-fired process steam generating plant and the purchase of electric power.

Inlet conditions: 850 psig/825°F
Exhaust pressure: 150 psig
Fuel price: \$1.40/MMBtu
Electricity price: 2.56¢/kWh
Annual standby by power cost:
\$50/kW for 15% of plant capacity

High fuel costs and low electric power rates also reduce the cost savings of a cogeneration plant. Because cogeneration enables industry to replace purchased electrical power with cogenerated power, a cogeneration plant is more attractive where the electric rate is high and the cost of industrial fuel is low. Although a cogeneration plant will consume more fuel than a boiler generating an equivalent amount of process steam to generate electric energy, the incremental fuel cost required to produce the cogenerated electric energy can be sufficiently low and the purchased electric cost sufficiently high, so that the ROI is attractive.

However, this analysis indicates that for most cogeneration configurations, given varying fuel prices and electric rates, the after-tax ROI does not generally meet the normal hurdle rates of the industries. Figures 3.6 and 3.7 illustrate the impact of different fuel prices and electric rates on the ROI.

Although energy costs for cogeneration vis à vis existing systems are essentially site specific, weighted energy costs were used for each industry based on the regional distributions of the industries, cost information obtained through industry discussions, and FEA regional projections. Table 3.16 presents energy cost assumptions for each industry.

High charges for utility standby service can produce low cost savings for a cogeneration plant. That is, the ROI of a cogeneration plant can be affected by the cost of standby electric power from a utility to enable continued operations, if the cogeneration plant is unable to generate the required electric power. The sensitivity of ROI to standby power requirements is illustrated in Figure 3.8.

The impact of purchased standby power depends on the amount of power required to maintain critical operations and the purchase price negotiated with the utility. Discussions with utilities and industry represéntives indicate that standby purchases are the exception rather than the rule. In most cases where standby power is purchased, however, the amount of power contracted is about 10 to 20% of total demand.

Business Risks

Management's perception of the risk and uncertainties surrounding a potential cogeneration project can affect the capital investment deci-

FIGURE 3.6: INCREMENTAL ROI OF COAL-FIRED COGENERATION PLANTS WITH VARYING FUEL PRICES AT EXISTING MANUFACTURING FACILITIES*

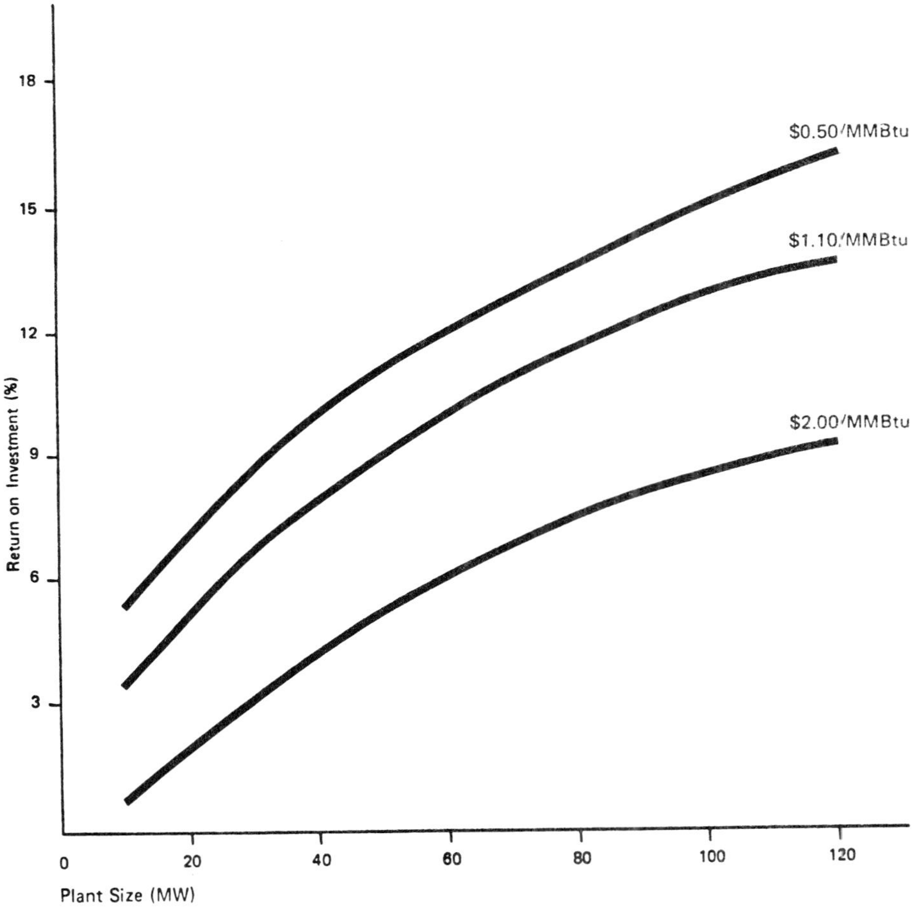

*The incremental ROI is that of a new coal-fired cogeneration plant compared to the continued use of an existing coal-fired process steam generating plant and the purchase of electric power.

Inlet conditions: 850 psig/825°F.
Exhaust pressure: 150 psig.
Annual capacity factor: 0.90.
2.7¢/kWh electric price assumed in all cases.
Annual standby power cost:
$50/kW for ½ of plant capacity.

FIGURE 3.7: INCREMENTAL ROI OF COAL-FIRED COGENERATION PLANTS WITH VARYING ELECTRIC PRICES AT EXISTING MANUFACTURING FACILITIES*

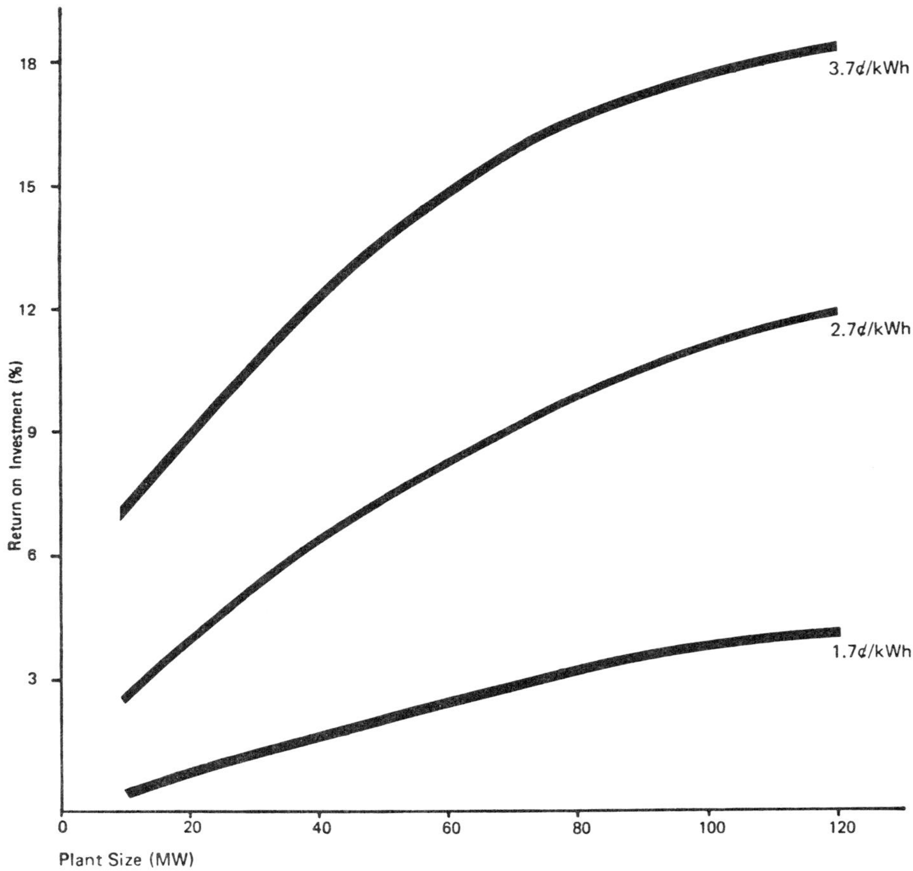

* The incremental ROI is that of a new coal-fired cogeneration plant compared to the continued use of an existing coal-fired process steam generating plant and the purchase of electric power.

Inlet conditions: 850 psig/825°F.
Exhaust pressure: 150 psig.
Annual capacity factor: 0.90.
Fuel price of $1.40/MMBtu assumed in all cases.
Annual standby power cost:
$50/kW for one-half of plant capacity.

FIGURE 3.8: INCREMENTAL ROI OF COAL-FIRED COGENERATION PLANTS WITH VARYING ANNUAL STANDBY CHARGES AT EXISTING MANUFACTURING FACILITIES*

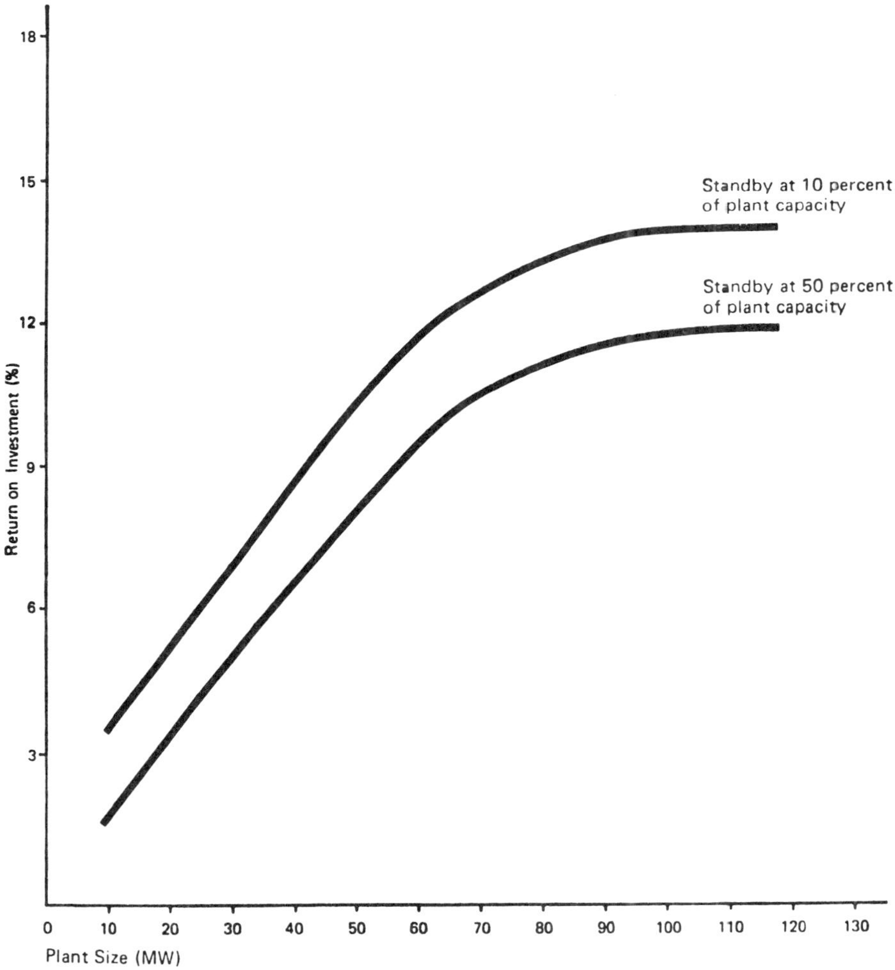

* The incremental ROI is that of a new coal-fired cogeneration plant compared to the continued use of an existing coal-fired process steam generating plant and the purchase of electric power.

Inlet condition: 850 psig/825° F.
Exhaust condition: 150 psig.
Annual capacity factor: 0.90.
Utility electric price: 2.7¢/kWh.
Fuel price: \$1.40/MMBtu.
Standby power cost: \$50/kW.

sion. The characteristics of a cogeneration project relate in part to the capital intensity of cogeneration investments and the long economic life of the project as well as previous experience with cogeneration. The risks and uncertainties of economic factors, environmental controls, and regulatory and institutional constraints diminish the acceptability of cogeneration projects. By causing the hurdle rate to be raised, these considerations reduce the potential for cogeneration development in all six industries by about 7 trillion Btu in 1976 and 97 trillion Btu by 1985 (see Table 3.12).

Although the specific managerial concerns that influence capital investment decision making are complex and unique to each company and each manufacturing plant site, it has been possible through discussions with industry decision makers to classify and estimate the major risks that limit cogeneration development. These were found to be capital availability, fuel availability and price, environmental policy and control equipment, and government regulation.

With only a few exceptions, management's decision criteria are explicitly or implicitly altered when cogeneration opportunities are being considered because of the above risk-related issues. In all industries considered, with the exception of the paper and pulp industry, the overall influence was to cause the requirement for a higher ROI. As Figure 3.9 indicates, the petroleum refining industry has the greatest increase in the median ROI hurdle rate; this high rate is primarily caused by capital constraints.

In contrast, management decision makers in the paper and pulp industry appear to place a higher priority (allow a lower ROI hurdle rate) on energy conservation projects that make companies more independent of electric utilities, allow a conversion to more secure fuels, or cut energy costs. This emphasis on energy-related projects is not confined exclusively to the paper and pulp industry; it is reflected in different degrees in the remaining five industries, particularly in the chemical industry where the risk-adjusted ROI hurdle rate is comparable to the normal ROI hurdle rate at low ROIs.

Capital Availability: The capital investment required for a cogeneration plant is considerable. Further, the value of the cogeneration plant relative to the manufacturing facility where it is to be located may be large; for example, the capital cost of a complete in-house cogeneration facility at an integrated paper mill may approach 60% of the value of the manufacturing plant.

FIGURE 3.9: CONGENERATION POTENTIAL THAT WILL DEVELOP UNDER RISK ADJUSTED HURDLE RATES*

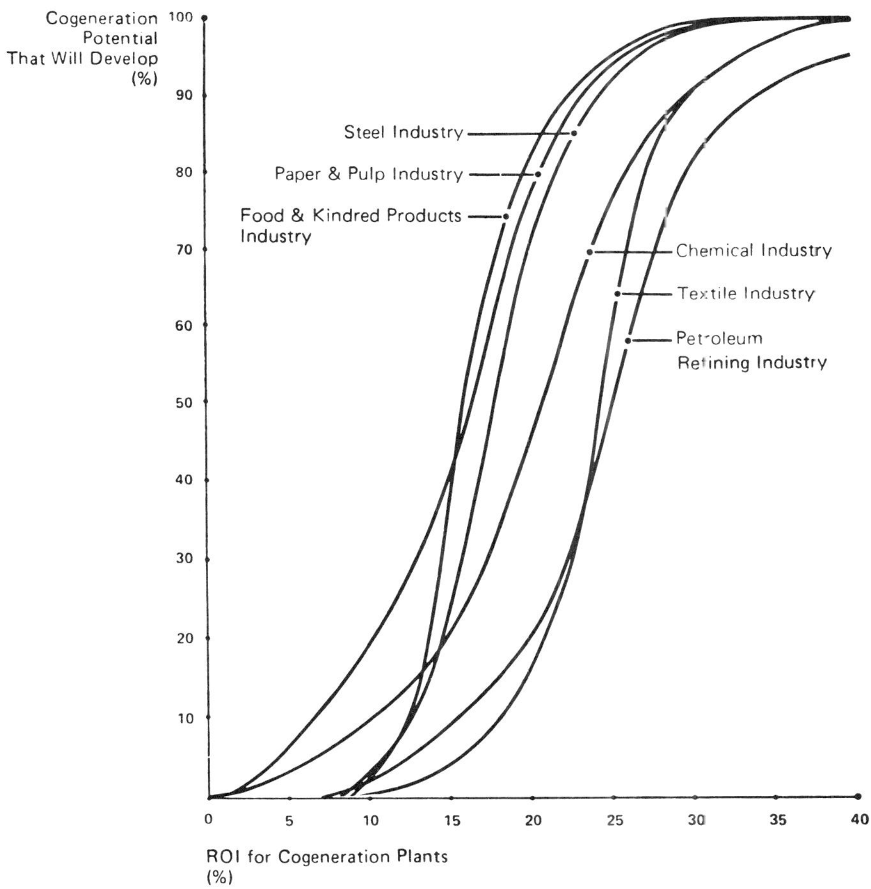

*Industry interview results showed that some companies do make energy conservation investments at low or negligible ROIs for policy, modernization, and other reasons.

Many companies do not have sufficient internally generated funds and their access to capital markets is limited. Therefore, the investment for a cogeneration plant could increase the company debt to an unacceptable level for the decision makers. For some companies, the important difference between market-oriented investments and cost saving projects further limits the acceptability of cogeneration-related investments.

Although the availability of capital can change over time for each company and industry, cogeneration development in the petroleum refining and steel industries appears to be most constrained by this factor.

Fuel Availability and Price: Managers are concerned that the fuel required for a cogeneration facility may become unavailable in the future, or may be unacceptably priced. Currently, there are significant supply and price questions associated with both petroleum products and coal. For instance, management is concerned about future curtailments in the supply of natural gas and oil, the potential effects of the deregulation of oil and gas prices, the lack of price stabilization of coal, and the future capability of the coal mining and transportation industries to meet demand.

In view of these uncertainties, some managers are reluctant to increase fuel demand by installing new or additional cogeneration facilities. To conserve energy, managers might plan substantial conversions from steam-powered to electrical-powered equipment, thus reducing the demand for steam and hence the potential for cogeneration.

The uncertainties and risks of fuel prices and availability affected the ROI hurdle rate of the chemical industry more than any other industry. This impact can be primarily attributed to the industry's dependence on foreign fuel markets (primarily in the mid-Atlantic area), and on natural gas for steam generation in the Southwest.

In contrast, little concern for future fuel prices and availability is evident in the paper industry, because of its dependence on diversified fuels including waste fuels. In the future, the industry generally expects and plans to have a heavier dependence on waste fuels, thereby further reducing its dependence on purchased fuels.

Environmental Policy and Control Equipment: Because of the additional fuel required for cogeneration, there will be an increase in localized air emissions. In cases in which two existing pollution sources (industrial process steam boiler and electric utility generation plant) are in sufficient proximity, cogeneration should result in a new reduction of emissions in the general area. This reduction, however, will depend on the fuels currently used by the utility to generate process steam and fuels proposed for the cogeneration plant.

The potential cost of environmental impacts has caused managers who are concerned about changes in state and federal pollution control requirements to demand a higher than normal ROI. In addition, there are significant technical and economic uncertainties associated with scrubbers (e.g., reliability and waste disposal problems) that cause managers to require a higher ROI.

Some managers have rejected or tabled cogeneration investments by requiring high ROI hurdle rates because they cannot find a way to reduce emissions sufficiently or to achieve offsetting reductions elsewhere. EPA nonattainment area policies have been the primary problems in the case of a few cogeneration opportunities. Even if the capability were available for meeting emission standards, the nonattainment policy contributes to a general uncertainty regarding future emission control requirements.

The two industries that appear to be most strongly influenced by these risks and uncertainties are the chemical and paper industries. For the chemical industry, this effect can be attributed in part to the fact that many of the manufacturing plants in this industry are located in environmentally sensitive areas. The paper industry, which is heavily dependent on diversified fuels, primarily waste fuels, is concerned about the adequacy of environmental control equipment to meet unpredictable environmental standards.

This uncertainty appears to have the least impact on the textile industry. This effect can be attributed to the concentration of manufacturing plants in nonsensitive environmental areas and the expected small dependence on coal for steam generation.

Government Regulation: Currently, the nature and extent of Federal Energy Regulatory Commission (FERC) and state public utility commission (PUC) jurisdiction over corporations that own or operate cogeneration facilities are not clearly defined. There are some situations in which the FERC and PUCs clearly would not have jurisdiction under current law (e.g., when an industrial cogeneration plant does not sell power and has no physical interconnection with the grid). In other cases, however, a clear regulatory policy does not exist. These uncertainties do not appear to be confined to any industry, but are indicative of the different historical experiences with on-site cogeneration and utility service among companies and specific manufacturing facilities.

For instance, management is concerned about the potential regulation of cogeneration facilities that parallel but do not transmit power to the utility grid. There is significant uncertainty over the prospect that industrial cogeneration plants could be required to deliver power to the grid to provide reserve or emergency capability for the utility and thus jeopardize manufacturing plant operations. In most instances, this uncertainty is represented by an increase in the required ROI.

Although the issue of potential regulation influenced the required ROI for cogeneration in all industries about equally, this risk was perceived to have the least effect on the hurdle rates. However, on the issue of selling power, the petroleum refining and chemical industries generally perceived the potential risks and uncertainties of regulation to be significant enough to remove this option from consideration. In contrast, the pulp and paper industry indicated that in some instances larger cogeneration plants would be installed if the wheeling of electric power were allowed.

SECTION 2
COGENERATION DEVELOPMENT WITH GOVERNMENT ACTION

Despite the technical and capital investment constraints already discussed, there is potential for government action to stimulate additional cogeneration development in the six manufacturing industries studied. By 1985, without government action, about 1,200 trillion Btu of process steam is financially unattractive, 1,100 trillion Btu of process steam is eliminated because of nontechnical factors (anticipated declining steam demand and excessive steam demand), and 2,400 trillion Btu of process steam is eliminated because of technical factors.

Analysis indicates that properly directed government programs could stimulate the development of between 300 and 900 trillion Btu of process steam cogeneration by 1985. As illustrated in Figure 3.10, development of this potential together with that existing in 1976 plus that which would develop without government action could result in the development of about 65% of the 3,500 trillion Btu which is technically suitable by 1985.

FIGURE 3.10: PROCESS STEAM FROM INDUSTRIAL COGENERATION FROM 1977 THROUGH 1985

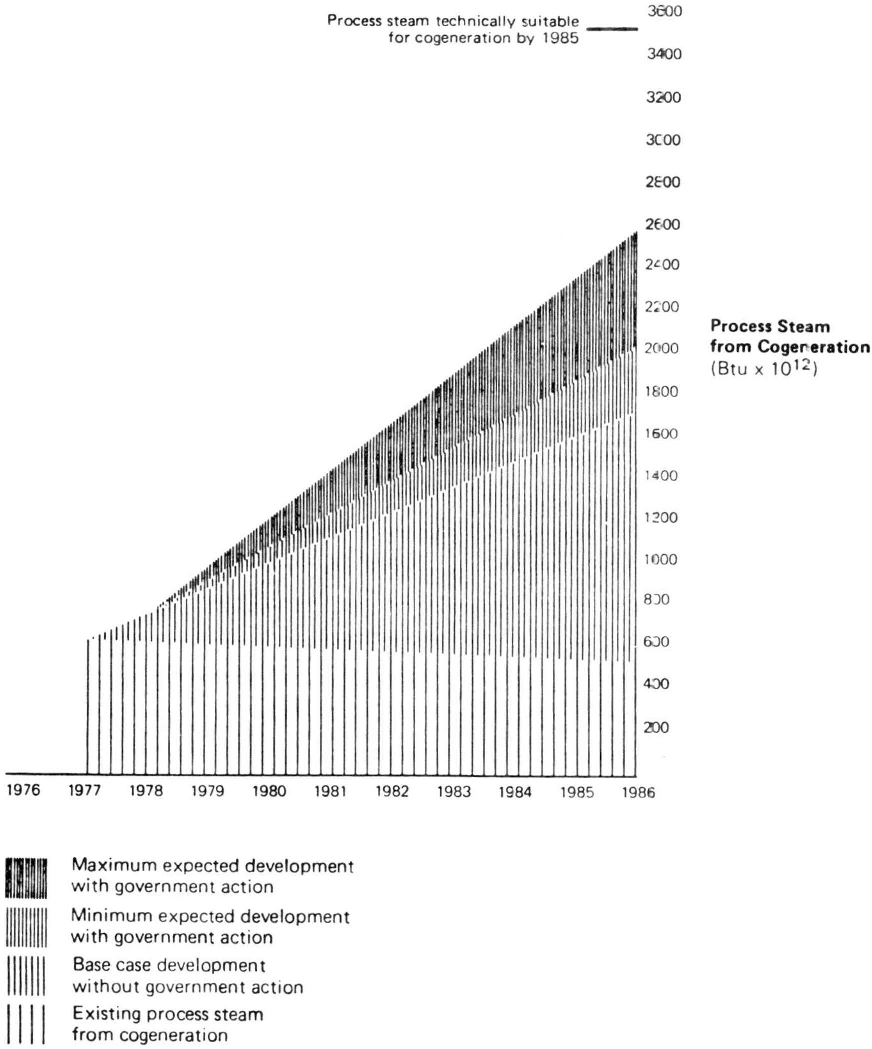

The additional cogeneration capacity attributed to these government programs is expected to be between 2,000 and 6,000 MW by 1985 in these industries; greater development could occur beyond 1985. The incremental energy savings realized by 1985 from this additional cogeneration capacity would be between 60 and 200 trillion Btu per year as illustrated in Figure 3.11. The major fuel saved would be oil (about 75% of the primary fossil fuel is saved in 1985). A summary of energy savings of six potential government programs is presented in Figure 3.12.

FIGURE 3.11: ENERGY SAVINGS FROM INDUSTRIAL COGENERATION WITH GOVERNMENT ACTION

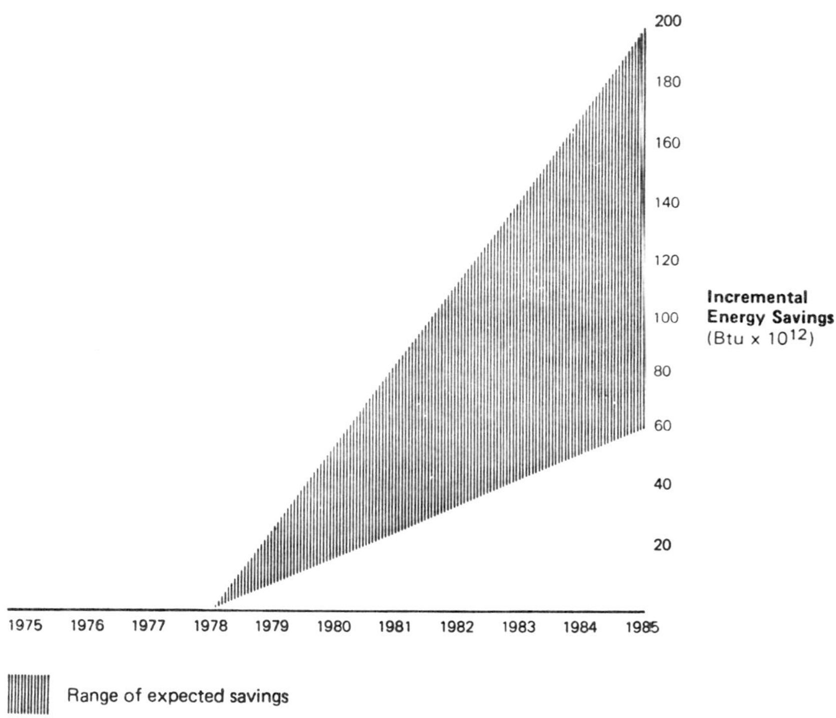

Federal programs should contain actions to provide immediate energy savings prior to 1985 and to initiate increased energy savings beyond 1985. Near-term energy savings (prior to 1985) can be achieved by government actions which mitigate the barriers to capital investments for on-site industrial cogeneration plants. Government actions focusing on technical barriers and utility involvement can produce additional savings beyond 1985.

FIGURE 3.12: ENERGY SAVINGS IN 1985 ATTRIBUTED TO POTENTIAL GOVERNMENT PROGRAMS

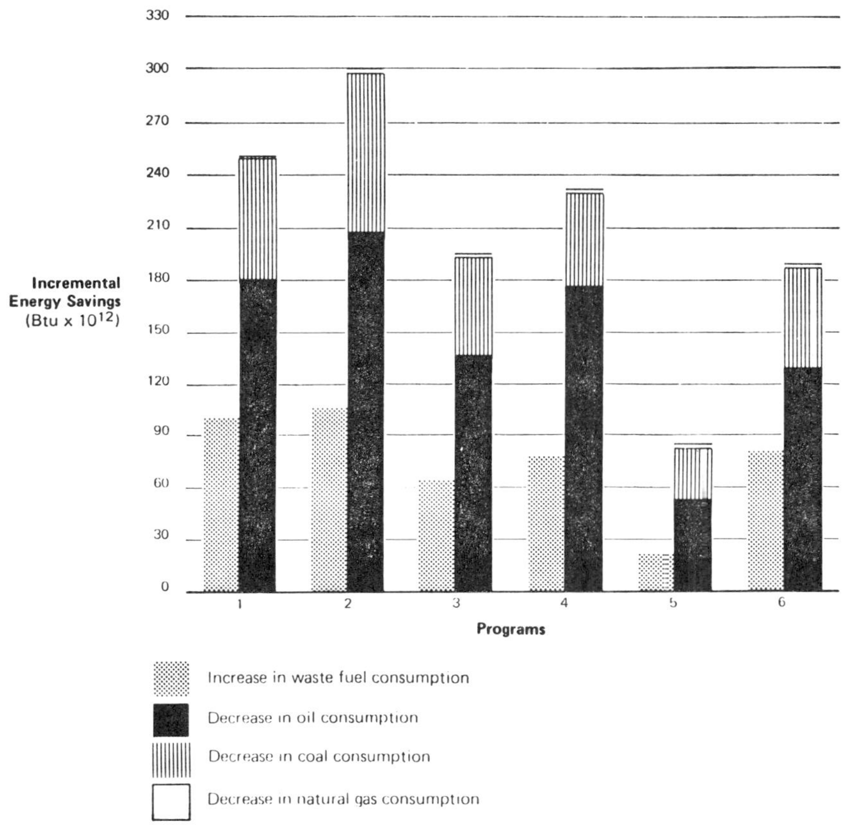

INDIVIDUAL GOVERNMENT ACTIONS

The federal government can use a wide range of actions to stimulate additional cogeneration development over the next decade. Information on the costs and benefits of each action is necessary to design an effective program to stimulate cogeneration. The impacts of 19 potential government actions on cogeneration development, on U.S. consumption of each major fuel, and on the federal budget have been examined. Generally, it was found that there was a significant potential for energy savings, but a wide spread in the cost effectiveness of these savings.

Actions Analyzed Quantitatively

Nine potential government actions were analyzed quantitatively in terms of their energy savings and costs to the federal government. The first three are financial incentives specifically designed to encourage cogeneration and their impact can be quantified with certainty. These actions are:

30% investment tax credit (ITC)

12-year depreciation option

Government loan guarantees.

The next two actions are regulatory, not financial. While resulting in significant energy savings, they have far-reaching effects beyond the scope of this study. They are not closely focused on cogeneration and can significantly affect the economy, the environment, or an established regulatory system. Consequently, the findings reported here should be interpreted with care. A decision to implement the actions must take into account not only their effectiveness in development of cogeneration potential but also their other impacts. The actions are:

Flat rates for electrical power

Relaxation to sulfur dioxide emission standards.

The five actions described above are the most "effective" energy savers. By "effective" is meant that they result in the greatest development of cogeneration and thus the largest energy savings. The four regulatory actions listed below, however, are the most "efficient" energy savers. By "efficient" is meant that they result in greater energy savings per dollar of net cost to the government. The actions are:

Equitable prices for required purchase of power from industrial cogeneration plants by utilities

Exemption from FERC and PUC jurisdiction for industrial cogeneration plants operating in parallel with, but not selling to, a utility

Exemption of industrial cogeneration plants from taxes on oil and gas use

Assignment of high priority to allocation of oil and gas for industrial cogeneration plants.

For each action analyzed quantitatively, effectiveness in terms of energy saving is a reliable measure. A great deal of confidence can be placed in the estimates of cogeneration that will develop because both steam and gas turbine configurations were analyzed to arrive at the estimates which were then confirmed in industry discussions.

The efficiency of each government action is a less certain measure than effectiveness and should be used primarily as a relative index for the actions. No quantitative criteria have yet been established to determine the price the government should be willing to pay to save 1 million Btu energy. The authors of this report believe that the cost to the government should not be measured in terms of the market price of energy, but should include measures of the social issues which are often not considered in private sector decision making. Such issues include environmental impacts, risks of shortages, and foreign policy pressures.

The cost estimates underlying the efficiency measure include both personnel-related government expenditures and net lost tax revenue to the treasury. For the proposed nonfinancial federal actions, the net lost revenues consist of the tax lost because of the existing 10% investment tax credit allowed for capital expenditures adjusted by the tax revenue gained through the higher corporate profits caused by the cogeneration plant. The financial incentives will cause additional lost tax revenues through either the additional 20% investment tax credit over the 10% already allowed or a depreciation tax shield, adjusted by greater corporate income taxes on the cogeneration development over the 1977–1985 period.

Although the first measure—effectiveness of the action in terms of energy saved—can be determined with the greatest degree of certainty, the amount of fuel savings resulting from each government action is difficult to quantify. Savings, it is believed, will take place primarily from the decreased use of baseload (and to a lesser extent, intermediate) generating units with the highest operating costs, either the oldest fossil-fuel units or the units burning the most expensive fuel (i.e., oil). The particular mix of fuels that will be displaced by cogeneration will be a function of unit-specific operating cost factors determined by each utility's economic dispatch system.

To estimate the reduction in fuel consumption by the utility, the following steps were taken:

> Utility generation was separated into baseload, intermediate, and peaking categories for each fuel in

each Electric Reliability Council (of which there are nine);

It was assumed that fuel would be displaced from each category in this sequence: natural gas first, oil second, and coal third;

It was assumed that because of the different industry electric demand patterns, cogeneration would displace intermediate utility generation in the food and textile industries and baseload generation in the other four industries.

Each of the nine government actions is discussed below in terms of its energy savings and cost (Table 3.17), its ultimate cogeneration development (see Table 3.18), and its savings by fuel type (see Table 3.19).

TABLE 3.18: COGENERATION DEVELOPMENT DUE TO INDIVIDUAL GOVERNMENT ACTIONS

Government Actions	Additional Cogenerated Process Steam (trillion Btu)	Cogenerated Electric Energy (billion kWh)	Electric Capacity of Cogeneration (MW)
30-percent ITC	416	20.0	2600
12-year depreciation	164	8.2	1050
Government loan guarantees	96	4.7	620
Electric utility flat rates*	293	14.1	1808
Relaxation of SO_2 emission standards**	126	5.9	752
Equitable selling prices†	54	3.1	407
Exemption from FERC, PUC regulation	35	1.6	220
Oil and gas use tax exemption	73	3.6	461
Oil and gas allocation priority	22	1.0	136

*This action assumes a 15-percent increase in the average costs of purchased electricity and does not include time-of-day considerations.

**This action can create severe local environmental and economic impacts not addressed in this study.

†This action makes no allowance for changes in cogeneration technology mixes (i.e., steam and gas turbines), it addresses only changes in plant sizes, including steam/electric ratios.

Thirty Percent Investment Tax Credit: The establishment of a 30% investment tax credit (ITC) is the most effective but the least efficient of the government actions analyzed quantitatively. Cumulative energy savings of 720 trillion Btu could be expected over the 1978 through 1985 period. However, the cumulative $2.5 billion cost is very high and the unit cost of $3.40 per million Btu saved is the highest of any action (with the exception of the 50% ITC for scrubbing equipment, which results in insignificant energy savings).

TABLE 3.17: IMPACTS OF GOVERNMENT ACTIONS

Government Actions	Net Energy Savings (trillion Btu) 1985	Net Energy Savings (trillion Btu) Cumulative 1978-1985[a]	Cumulative Cost to Government 1978-1985[a] (million dollars)	Cumulative Cost to Government per Million Btu Saved (dollars)	Energy Savings Mix in 1985: Oil as a Percent of Fuels Saved[a]	Energy Savings Mix in 1985: Ratio of Oil Savings to Net Energy Savings
30-percent ITC	94	720	2500	3.40	70	1.2
12-year depreciation	38	295	800	2.70	65	1.1
Government loan guarantees	21	160	45	0.25	80	1.4
Flat rate structure[b]	65	480	290	0.60	70	1.2
SO_2 emissions relaxation[c]	24	170	(120)[d]	(0.70)	105	1.6
Equitable prices[e]	13	100	160	1.60	100	1.0
Exemption from FERC/PUC jurisdiction	8	60	8	0.15	70	1.2
Exemption from use tax	16	120	85	0.70	70	1.2
Allocation priority	5	40	5	0.10	–[f]	–[f]

a Figures have been rounded.
b This action assumes a 15-percent increase in the average costs of purchased electricity and does not include time of day considerations.
c This action potentially can create severe local environmental and economic impacts not addressed in this study.
d Numbers in parentheses indicate a net revenue to the government.
e This action makes no allowance for changes in cogeneration technology mixes (i.e., steam and gas turbines), but addresses only changes in plant sizes, including steam/electric ratios.
f This action does not produce a net savings of oil.

TABLE 3.19: NET ANNUAL ENERGY SAVINGS IN 1985 BY FUEL TYPE THROUGH GOVERNMENT STIMULATED COGENERATION (trillion Btu)

Individual Government Actions	Coal			Oil			Gas			Waste Fuels			Net Total		
	Cogeneration Increase	Utility Reduction	Net Savings	Cogeneration Increase	Utility Reduction	Net Savings	Cogeneration Increase	Utility Reduction	Net Savings	Cogeneration Increase	Utility Reduction	Net Savings	Cogeneration Increase	Utility Reduction	Net Savings
30-percent ITC	40.3	(78.9)	(38.6)	16.5	(128.0)	(111.5)	–	(1.3)	(1.3)	57.9	–	57.9	**114.7**	**(208.2)**	**(93.5)**
12-year depreciation	12.2	(34.3)	(22.1)	8.1	(50.3)	(42.3)	–	(0.4)	(0.4)	27.0	–	27.0	**47.3**	**(85.0)**	**(37.7)**
Loan guarantees	8.4	(15.1)	(6.7)	4.6	(33.0)	(28.4)	–	(0.7)	(0.7)	15.2	–	15.2	**28.2**	**(48.7)**	**(20.6)**
Electric utility flat rates*	27.6	(56.2)	(28.6)	12.7	(89.1)	(76.4)	–	(0.8)	(0.8)	40.8	–	40.8	**81.1**	**(146.1)**	**(65.0)**
Relaxation of SO_2 emission standards	25.0	(23.2)	1.8	0.0	(38.1)	(38.1)	–	(0.2)	(0.2)	12.3	–	12.3	**37.3**	**(61.5)**	**(24.2)**
Equitable selling prices**	12.2	(11.9)	0.3	6.9	(20.2)	(13.3)	–	(0.3)	(0.3)	0.0	–	0.0	**19.1**	**(32.4)**	**(13.3)**
Exemption from FERC, PUC regulation	3.0	(6.8)	(3.8)	0.3	(10.0)	(9.7)	–	(0.2)	(0.2)	5.5	–	5.5	**8.8**	**(17.0)**	**(8.2)**
Oil and gas use tax exemption	5.5	(14.1)	(8.6)	2.8	(23.0)	(20.2)	–	(0.2)	(0.2)	12.8	–	12.8	**21.1**	**(37.3)**	**(16.2)**
Oil and gas allocation priority	0.0	(4.7)	(4.7)	6.0	(5.7)	0.3	–	(0.2)	(0.2)	0.0	–	0.0	**6.0**	**(10.6)**	**(4.6)**

*This estimate is based on an assumed 15-percent increase in the cost of purchased electricity and does not include time-of-day considerations.

**This is a conservative estimate since no allowance was made for changes in cogeneration technology mixes, but only a change in plant sizes.

This tax credit is purely a financial incentive with no regulatory or technical component to it. Specifically, the action must be taken by Congress and requires a federal tax code provision adding to the existing cogeneration ITC of 10% an additional 20% for new cogeneration equipment. The government must define cogeneration equipment so that the term excludes, to the extent possible, investments which would have been made even if cogeneration had not been involved (e.g., low-pressure process steam boilers). For these calculations, the additional 20% ITC was applied to the incremental cogeneration investment. This incremental investment included the whole cost of a new higher-pressure boiler in all cogeneration applications except where mandatory coal conversion was assumed, and included only the incremental cost of a high-pressure boiler (over and above a boiler for process steam) for all new development and coal conversion applications.

A 30% ITC would increase cogeneration development by improving the financial returns and increasing the percentage of technically suitable steam developed in those instances where cogeneration is already financially attractive. For example, investment in a 30 MW, new development, coal-fired, steam turbine cogeneration plant (under technical and energy price conditions assumed for the chemical industry) would have an ROI of 20.5% with the tax credit; this compares to an ROI of 16.2% without the credit.

For all six industries, only 35% of the total steam technically suitable for cogeneration development in 1985 would have sufficiently attractive ROIs to cause investment in cogeneration over the 1977–1985 period without government action. With a 30% ITC, however, 48% of the steam becomes financially attractive. Thus, by 1985, the incremental cogeneration development due to the tax credit is 416 trillion Btu of steam energy and 20 billion kWh of electric energy; electric generating capacity increases by 2,600 MW.

To cogenerate additional electricity, industry would burn more coal, oil, and waste fuel, and would recover more waste heat. Since this additional industrial generation would displace an equivalent amount from the electric utilities' systems, the savings of coal, oil, natural gas, or nuclear fuel would be effected at the utilities. The net effect would be a reduction in U.S. energy consumption by 1985 of 94 trillion Btu per year. This reduction would result from a 39 trillion Btu reduction in coal use, a 112 trillion Btu reduction in oil use, a 1 trillion Btu reduction in natural gas use, and a 58 trillion Btu increase in the use of process residual (waste) fuel.

The primary cost associated with the 30% ITC would be $2.5 billion (1976 dollars) of lost revenue to the treasury over the eight-year period from 1978 through 1985. Spreading this cost over the 720 trillion Btu of cumulative energy savings would result in a cost to the government of $3.40 per million Btu. Thus, of all those actions analyzed quantitatively, this is the most costly and the least efficient in saving energy.

Twelve Year Depreciation Option: The establishment of a 12 year depreciation option for industrial cogeneration equipment is one of the most effective of the actions analyzed quantitatively but is a relatively inefficient energy saver. Cumulative energy savings of 295 trillion Btu could be expected over the 1978 through 1985 period. However, the cost of $2.70 per million Btu saved would represent a total cost of $0.8 billion. It is purely a financial incentive with no regulatory or technical component to it. Specifically, the action must be taken by Congress and would require a federal tax code provision.

The proposed depreciation schedule is shorter than is now allowed under federal tax codes (22½ years on equipment). It would increase cogeneration development by improving financial returns and increasing the percentage of technically suitable steam developed in those instances where cogeneration is financially attractive. For example, analysis indicates that investment in a 30 MW, new development, coal-fired, steam turbine cogeneration plant (under technical and energy price conditions assumed for the chemical industry) would have an ROI of 17.4% with this depreciation schedule; this compares to an ROI of 16.2% without the action.

For all six industries, only 35% of the total steam technically suitable for cogeneration development in 1985 will have sufficiently attractive ROIs. With this action, however, about 40% of the steam would become financially attractive over the 1978–1985 period. The incremental 5% increase in cogeneration development resulting from this action represents 164 trillion Btu of steam energy cogeneration, 8.2 billion kWh of cogenerated electric energy, and incremental electric generating capacity of about 1,050 MW.

The net energy savings effect would be a reduction in U.S. energy consumption by 1985 of 37.8 trillion Btu per year. This savings would result from a 22 trillion Btu reduction in coal use, a 42 trillion Btu reduction in oil use, a 0.4 trillion Btu reduction in natural gas use, and a 27 trillion Btu increase in the use of waste fuel.

The primary cost associated with this accelerated depreciation would be the $800 million in revenue lost to the treasury over the eight-year period from 1978 through 1985. Spreading this cost over the 295 trillion Btu of cumulative energy savings would result in a cost to the government of $2.70 per million Btu. Thus, while this action would be more efficient than the investment tax credit, it would be less efficient, or more costly, than are the other actions analyzed quantitatively.

Government Loan Guarantees: The establishment of a loan guarantee program is one of the most effective of the actions analyzed quantitatively. While cumulative energy savings could be expected to be about 160 trillion Btu over the 1978 through 1985 period, the cumulative $45 million cost over the period is low. The cost of about $0.25 per million Btu saved would make this action an efficient energy saver.

This loan guarantee program would increase cogeneration development primarily in small companies or subsidiaries by reducing the risk and thus the ROI hurdle rates. The amount of technically suitable steam developed would probably be small, since only the steel and petroleum refining industries anticipate severe capital constraints. For all six industries, the incremental investment in cogeneration with government loan guarantees is only 3% (that is, 38% of the technically suitable steam would be cogenerated). Therefore, the incremental increase in cogeneration development that would result from this action represents 96 trillion Btu of steam energy cogeneration, 4.7 billion kWh of electric energy cogeneration, and an electric generating capacity of about 620 MW.

The net energy savings effect would be a reduction in 1985 U.S. energy consumption of 21 trillion Btu. This savings would result from a 7 trillion Btu reduction in coal use, a 28 trillion Btu reduction in oil use, a 1 trillion Btu reduction in natural gas use, and a 15 trillion Btu increase in the use of waste fuel.

The primary cost associated with this action would be $45 million over the eight-year period from 1978 through 1985. Spreading this cost over the 160 trillion Btu of cumulative energy savings would result in a cost to the government of about $0.25 per million Btu. Thus, this action is one of the more efficient and least costly actions proposed in this study.

Flat Rates for Electrical Power: It is estimated that if flat rates for electrical power purchased by industrial customers resulted in an average increase of 15% (a benchmark estimate) in the price of electricity purchased from utilities by large industrial users, then about 300 trillion Btu of additional steam cogeneration would be developed. The establishment of flat rates would therefore be the second most effective action in terms of increasing cogeneration (after the 30% investment tax credit). The cumulative $290 million cost to the government attributable to this action over the 1978–1985 period because of a net loss of tax revenues and administrative costs, and the energy savings of 500 trillion Btu would result in an average cost to the government of $0.60 per million Btu of energy saved. Compared to the cost of the 30% ITC—$3.45 per million Btu of energy saved—this is a highly cost-efficient action.

The establishment of flat rates for electricity would result in a significant increase in the electricity bill for large baseload customers. In effect, the new flat rates should be established for a class of customers so that the net return to the utility for that class of customer remains unchanged after a consumption adjustment. The new flat rates would therefore be set at a level between the high cost per kWh paid by small users and the low cost per kWh paid by large users. Thus, customers with low consumption would be paying less and would be induced to increase their electricity consumption. On the other hand, customers with greater consumption would be paying more and would be induced to decrease their consumption of electricity and/or to shift their dollar allocation for energy toward in-house production.

It is assumed here, however, that large industrial electricity users cannot significantly reduce their power requirements. This is because manufacturing operations generally do not have the flexibility to decrease electric manufacturing processes which require large capital investments. Thus, the establishment of flat rates should encourage industry investments in cogeneration plants.

The assumed 15% increase in the price of electricity resulting from flat rates would encourage industries to cogenerate their own electric power. For example, an investment in a 30 MW, new development, coal-fired, steam turbine cogeneration plant (under technical and energy price conditions assumed for the chemical industry) would have an ROI of 19.0% with flat rates compared to an ROI of 16.2% with the present rates.

Cogeneration development by 1985 would increase from 35% without the action to 44% with the flat rates, given the assumption previously indicated. The amount of increase in cogeneration development that would result above the base-case development is 293 trillion Btu of steam cogeneration, 14.1 billion kWh of electric energy cogeneration, and an electric generating capacity of 1,808 MW.

To cogenerate additional electricity, industry must burn more coal, oil, and waste fuels and recover more waste heat. However, since this generation would reduce electric utility fuel consumption, less coal, oil, natural gas, or nuclear fuel would be required. The net effect would be a reduction in U.S. annual energy consumption of 65 trillion Btu of 1985. This would result from a 29 trillion Btu reduction in coal use, a 76 trillion Btu reduction in oil use, a 1 trillion Btu reduction in natural gas use, and a 41 trillion Btu increase in the use of waste fuel.

The primary cost of $290 million, attributable to the increase in cogeneration capacity, would be due to a net loss in tax revenue and the administrative costs required. Spreading this cost over the 500 trillion Btu of cumulative energy savings would result in a cost to the government of $0.60 per million Btu saved.

The establishment of flat rates would have a significant effect on all electricity users. Opposition could be expected from all those who fear that a new rate structure would not be to their immediate advantage. The following negative effects would have to be minimized:

> Flat rates of electrical power could reduce the cost to the small users who might increase their consumption and fail to conserve energy.
>
> Flat rates can run counter to the current emphasis of covering all costs incurred in producing and delivering energy. Thus, flat rates could result in a suboptimal overall allocation of scarce resources.

Relaxation of Sulfur Dioxide Emission Standards: The relaxation of sulfur dioxide (SO_2) emission standards is one of the more effective government actions that could be taken, because cumulative energy savings of 170 trillion Btu could be anticipated from 1978 through 1985. There would actually be a net revenue over the period of $120 million attributed to additional tax revenue obtained from companies installing coal-fired cogeneration plants that produce high returns.

This action would improve the expected incremental return on investment by removing the requirement for installation of desulfurization equipment. Specifically, the relaxation of sulfur emission standards from 1.2 pounds to 3.0 pounds per million Btu would allow operators of cogeneration plants to burn this higher sulfur fuel without the installation of flue-gas desulfurization equipment or "scrubbers." Because of existing government policy concerning the conversion of oil- and gas-fired industrial operations to coal, for the purpose of analysis, it is assumed that this action would deal exclusively with the utilization of high-sulfur coal. This action would allow coal mined in Coal Production Districts (CPDs) 8 (West Virginia, Virginia, Kentucky, Tennessee) and 13 (Alabama) to be burned without scrubbers.

Under this action, operators of coal-fired industrial boilers would still be required to burn compliance coal or use scrubbers to meet air emission standards. Regulations regarding other polluting emissions would remain unchanged for both cogeneration plants and industrial boilers.

The effect of this government action would be to decrease the incremental costs associated with coal-fired cogeneration operation, thereby making cogeneration plants the more financially attractive alternative. The financial savings would result from an assumed scrubber cost of $0.30 per million Btu that operators of coal-fired cogeneration plants will not have to pay.

This effective shift in fuel cost would cause an average increase in the after-tax ROI of 3 to 5 percentage points for coal retrofit and new coal facilities. Cogeneration development by 1985 would increase from 35% of the technically suitable steam without the action to 39% with this action. The increase in cogeneration development due to this action would be reflected by increases of 126 trillion Btu of cogenerated steam, 5.9 billion kWh of electric energy cogeneration, and 752 MW of cogeneration capacity.

To cogenerate additional electric energy, industry would install coal-fired and high-sulfur, waste fuel-fired cogeneration plants. The increase in fuel consumption for industrial cogeneration would reduce utility fuel consumption; less coal, oil, natural gas, or nuclear fuel would be required. The net effect would be a reduction in U.S. annual energy consumption of 24 trillion Btu by 1985. This would result from a 2 trillion Btu increase in coal use, a 38.1 trillion Btu reduction in oil use, a 0.2 trillion Btu reduction in natural gas use, and a 12.3 trillion Btu increase in the use of waste fuel.

A net cumulative revenue of approximately $120 million dollars over the period is attributable to the increased revenue from companies which would install cogeneration plants using high-sulfur fuels instead of more expensive low-sulfur fuels. Some managers who would install oil-fired cogeneration plants otherwise, would consider coal-fired cogeneration plants the investment with the greater return. These greater returns to industry would result in a significant increase in federal income tax revenue. This increased tax revenue is offset somewhat by the loss in tax revenue (investment tax credit) from the additional development and the costs required to administrate compliance with this action.

The most significant costs of this action are environmental. By 1985, an incremental 700 thousand tons of sulfur per year would be emitted to the atmosphere because of this action which would allow the burning of fuels with a sulfur content up to 3 pounds per million Btu in cogeneration facilities. This net increase would be the difference between the total sulfur emissions of a cogeneration unit burning a higher sulfur fuel without a scrubber and the total emissions from an industrial boiler and a utility boiler. The fuel/technology distributions for the expected 1985 development were used for the estimates presented. This impact could be attributed to both cogeneration development stimulated through the implementation of this action and development that would have occurred without this action. Existing cogeneration facilities would also come under this action. In this study, it was assumed that all facilities using low-sulfur coal would shift to high-sulfur coal. Careful consideration must be given to localized impacts of the relaxed standards before a decision should be made as to the efficacy of the action.

Equitable Prices for Power Purchased by Utilities: Government action to ensure equitable prices for power purchased from industrial cogeneration by utilities would result in moderate cumulative energy savings of 100 trillion Btu for 1978 through 1985, and the cost to the government would be relatively high at $1.60 per million Btu saved. In addition, there is more uncertainty concerning these savings and cost estimates than there is for the other actions analyzed quantitatively, due to the substantial regulatory issues now unresolved.

Prices at which the sale of cogenerated power would become financially attractive to an industrial plant were estimated. The economic benefit of selling excess power is the difference between the marginal cost of producing excess power and the price at which the cogenerated power can be sold. The marginal cost of industrial cogeneration is lower when sale of power is planned because of the economies of scale realized when plant size is increased. In instances where the steam/electric demand ratio precluded sizing cogeneration plants to match the steam demand without this action (see Section 1), cogeneration opportunities were considered where the cogeneration plant èqualed the baseload steam demand, thereby allowing the production of excess electricity. For these cases, it was found that when the sale price for the cogenerated power is about 10 to 20% below the purchase price for utility power, a cogeneration plant with capacity to sell power has approximately the same ROI as a plant that can only match industrial power needs.

Because some added risk is associated with selling power (e.g., long-term nature of sale contracts creates risks for an industrial plant that needs flexibility to react to changing market conditions), a higher ROI is generally required. For sale of power to be financially attractive, an industrial plant will require a selling price for cogenerated power that is roughly 10% below the plant's purchase price for utility-generated power.

In practice, regulatory agencies will address issues not included in this assessment. To establish equitable prices, however, regulators must compare the costs of power from industrial cogeneration with those from a utility's other supply alternatives.

Roughly 407 MW of cogeneration capacity was estimated to result from this action. This estimate is based on consideration of only those plant size increases that enable the sale of excess power and produce ROIs comparable to cogeneration plants sized for industrial plant needs only. The increase in steam energy cogeneration would be 54 trillion Btu in 1985, and the increase in electric energy cogeneration would be 3.1 billion kWh.

Energy savings in 1985 could be expected to be 13 trillion Btu, while cumulative savings for 1978 through 1985 would be 100 trillion Btu.

The total cost of this action would be $160 million. Its unit cost of $1.60 per million Btu saved is relatively high, making it a somewhat inefficient incentive.

Exemption from FERC and PUC Jurisdiction: Government action to exempt from electric utility regulation industrial cogeneration plants only paralleling and not selling power would have relatively low effectiveness. Cumulative energy savings of only 60 trillion Btu could be expected over the 1978–1985 period. However, the cumulative $8 million net cost over the period would be low, and the resulting $0.15 cost per million Btu saved would make this action more efficient than all but one other action.

This action would be the most effective response to the expressed concern by management that an industrial plant might become subject to the authority of the state public utility commissions if there were a parallel interconnection with the grid. (Parallel interconnections, desirable for reasons of security and stability, are live connections with the grid that allow a two-way plan of electric power, the purpose of which is to permit the power purchased from the utility to supplement the output of the industrial plant's cogeneration unit. In fact, this mechanism also permits power produced by the industrial plant to flow into the utility's grid.) It would require that FERC and state PUCs relinquish all jurisdiction over industrial cogeneration plants, except perhaps during a declared state of emergency, but would in no way restrict federal or state authority over the sale of power from industrial cogeneration plants.

This exemption would increase cogeneration investment by reducing the ROI required by companies for cogeneration, thus increasing the percentage of technically suitable steam which is financially attractive. For example, stimulated investments in new development, coal-fired, steam turbine cogeneration plants would increase cogenerated steam by an increment of 3% in the steel industry to 7% in the food industry over that developed in each industry without government action.

This action would increase the amount of steam technically suitable for cogeneration development in 1985 by only 1%. Therefore, the incremental increase in cogeneration development as a result of this action would be 35 trillion Btu of steam energy cogeneration, 1.6 billion kWh of electric energy cogeneration, and an electric generating capacity of about 220 MW.

The energy savings effect would be a reduction in 1985 U.S. energy consumption of 8 trillion Btu, resulting from a 4 trillion Btu reduction in coal use, a 10 trillion Btu reduction in oil use, a 0.2 trillion Btu reduction in natural gas use, and a 6 trillion Btu increase in the use of waste fuel, in 1985.

The net cost to the government associated with this action would be

$8 million of personnel-related expenditures and lost revenue to the Treasury from 1978 through 1985. Spreading this cost over the 60 trillion Btu of cumulative energy savings indicates a cost per million Btu saved of $0.15. Thus, this action is a more efficient way to save energy than all but one other. It is one of the two actions (the other being the relaxation of sulfur dioxide emission standards) being considered that would actually result in tax revenues for the government. This projected cumulative revenue is attributed to the greater income tax revenues versus the costs of a 10% investment tax credit. This net government revenue would only partially offset the high administrative costs associated with the implementation of this action.

Exemption from Tax on Oil and Gas Use: Exemption from oil and gas use taxes is one of the more cost-effective actions analyzed quantitatively. Although cumulative energy savings of only 120 trillion Btu could be expected over the 1978 through 1985 period, the $85 million net cost over the period would be comparatively low. The resulting cost of $0.70 per million Btu saved makes this action an efficient energy saver.

A tax on oil and gas use is currently under consideration in the Congress. This proposed tax on industry beginning in 1979 and on utilities beginning in 1983 has not been included in this base-case analysis. There would be no direct effect from the imposition of the tax if it included an exemption for cogeneration. In any case, it is expected that the impact of the proposed tax on electric utilities would raise electric rates by about 2 to 5%, depending on the anticipated fuel-consuming characteristics of the utilities at that time. Analysis indicates that a 3% increase in industrial electric rates nationally is reasonable, and that cogeneration development would be stimulated.

This action will increase cogeneration development by modestly improving cogeneration ROIs and increasing the percentage of technically suitable steam where cogeneration is financially attractive. For example, an investment in a 30 MW, new development, coal-fired, steam turbine cogeneration plant (under technical and energy price conditions assumed for the chemical industry) would have less than one percentage point change in the return on investment. The incremental increase in total steam technically suitable for cogeneration development in 1985 as a result of this action would be 2%, which represents an increase in the cogeneration development of 73 trillion Btu of steam energy cogeneration, 3.6 billion kWh of electric energy cogeneration, and an electric generating capacity of 461 MW.

The energy saving effect of this action would be a reduction in 1985 U.S. energy consumption of 16 trillion Btu, resulting from a 9 trillion Btu reduction in coal use, a 20 trillion Btu reduction in natural gas use, and a 13 trillion Btu increase in the use of waste fuel.

The primary cost associated with this action would be $85 million of lost revenue to the treasury over the eight-year period from 1978 through 1985. Spreading this cost over the 120 trillion Btu of cumulative energy savings indicates a federal cost of $0.70 per million Btu. Thus, this action would be relatively efficient way to promote energy savings.

Oil and Natural Gas Allocation Priority: Government action to assign a high priority to industrial cogeneration facilities in oil and gas allocation would have low effectiveness. Cumulative energy savings of only 40 trillion Btu could be expected over the 1978 through 1985 period as a result of this action. However, the cumulative cost over the period is low, and the resulting cost per million Btu saved ($0.10) makes this action very efficient.

The standby petroleum allocation program for use in the event of an embargo or other oil shortage was not perceived as a significant risk by the participants in the discussions which were held with industrial representatives, although there is widespread uneasiness about the general long-term availability of oil. Therefore, the opportunity to receive a high allocation priority for cogeneration plants using oil would affect the ROIs required for cogeneration investments only slightly. While natural gas curtailments are of much greater concern to managers with respect to their current and near-term operations, it is expected that very little natural gas would be used as a fuel in cogeneration. Furthermore, it is uncertain how significant an improvement in curtailment priorities the FERC and state PUCs would be willing to make for power and steam generation, when alternate fuels are readily available. Relatively small adjustments in industry's risk assessments have therefore been taken in the analysis of this action. Nevertheless, if natural gas were made free of curtailments for use in cogeneration gas turbines where the electrical output (and therefore the energy savings) is high, the effect of the action would be greater.

The increase in cogeneration development resulting from this action, above the development expected without government action would be 22 trillion Btu of steam energy cogeneration, 1 billion kWh of electric energy cogeneration, and an electric generating capacity of 136 MW in 1985.

A cumulative energy savings of 40 trillion Btu would result from 1978 through 1985 from this action. The net effect would be a reduction in 1985 U.S. energy consumption of about 5 trillion Btu, resulting from a 5 trillion Btu reduction in coal use, a 0.3 trillion Btu increase in oil use, and a 0.2 trillion Btu reduction in natural gas use in 1985.

The primary cost associated with this action would be $5 million of personnel-related expenditures and lost revenue from 1978 through 1985. Spreading this cost over the 40 trillion Btu of cumulative energy savings indicates a federal cost of $0.10 per million Btu saved. Thus, this action would be a more efficient inducement to save energy than any other action considered.

Actions Analyzed Qualitatively

Several additional government actions could induce energy savings, but they were not subjected to a quantitative analysis of these savings for this report. Other actions have not yet been defined at either the national or the utility service area levels so as to permit quantitative analysis. Each of the following potential actions will be discussed qualitatively: establishment of time-of-use electric rates for industry; clarification of FERC and PUC jurisdiction over the sale of cogenerated power; intervention in state regulatory proceedings; and technical assistance and policy studies for state PUCs.

Several other actions were not quantified because data were insufficiently conclusive and/or the impacts of the action did not appear sufficiently large to justify rigorous treatment. These were: federal information clearinghouse; antitrust exemption for joint ventures; exemption from mandatory coal conversion; support for demonstration projects and technical assistance; and operator training grants.

Establishment of Time-of-Use Electric Rates for Industry: Mandatory time-of-use electric rates for industrial customers will not necessarily stimulate industrial cogeneration. The impact of this measure depends on the consumption characteristics of the industrial customer and the flexibility of industrial customers to alter those characterictics. Although site-specific studies are required to accurately assess the impacts of industrial time-of-use rates on cogeneration development, some general conclusions can be drawn.

Mandatory time-of-use industrial rates should cause a decrease in the cost of electricity to baseload customers and an increase in costs to nonbaseload customers. For any given industrial customer class defined by electric power consumption levels, the net return to the utility under time-of-use rates, should remain unchanged in the short term. Under the time-of-use rates, the purpose is to charge high rates during the peak periods to reflect incurred utility costs more accurately. Because customer classes generally reflect different consumption patterns, the customer who requires power only during the utility's peak hours would generally be obliged to pay more than before and the customer consuming constant levels of electricity over a 24-hour period should pay less than before. If purchased electric energy represents a large portion of overall production costs at a manufacturing facility and plant operations are inflexible, then cogeneration may be an attractive investment, particularly for nonbaseload customers.

The baseload customer, the largest customer class, is not expected to incur greater costs for purchased electricity under a mandatory time-of-use rate. This type of customer would be motivated, in any case, to change his consumption pattern so as to minimize his purchased electricity charges. For these customers, the action would not encourage cogeneration investments.

The direct costs to the government would be only those associated with the technical assistance and other support to state PUCs to establish guidelines and regulations, which have already been discussed.

Besides the effects on cogeneration of time-of-use rates, the utilities' load factor would increase, therefore decreasing the oil and gas (expensive fuels) consumed to meet a peaking demand for electricity. The capital requirements for building additional peaking capacity could be significantly reduced.

In addition, time-of-use rates track costs better and are closer to a marginal pricing system. This means the rates should induce optimal allocation of scarce resources. Additional costs and problems associated with time-of-use rates could include:

> Companies with more flexibility to adjust their consumption pattern will benefit from the new prices at the expense of either the utilities, in the short term, or the nonflexible users over a longer term. It may therefore be difficult to maintain a "fair system."
>
> There are significant administrative costs as well as costs for installing metering equipment.

> New peaks might be created if the price differential provides incentives which are large enough to cause significant shifts in demand to off-peak periods.
>
> To the extent that average cost of electricity would decrease because of increased efficiency and load factors, cogeneration will become less attractive.

Clarification of FERC/PUC Jurisdiction Over the Sale of Cogenerated Power: Under this action, the Federal Energy Regulatory Commission would hold rule-making proceedings which would specify the conditions under which FERC would have jurisdiction over the sale of cogenerated power by industry. The extent of the regulations would also be determined. In addition, Congress would pass legislation specifying the maximum extent of state public utility commission jurisdiction over the sale of cogenerated power and requiring state PUCs to hold their own rulemaking proceedings to clarify their regulatory authority in this area. Finally, FERC would, upon completion of these state and federal proceedings, prepare and make available a report documenting all potential jurisdiction over power sale by industrial and joint-venture cogeneration plants.

This action was examined because of industry's expressed concern over the regulatory consequences of selling power. Consequently, the impact of the action on cogeneration would depend primarily on the financial attractiveness of the sale of power, which, in turn, depends on the implementation of other actions such as the requirement of equitable sale prices for cogenerated power. At present, it is generally more attractive for industries to cogenerate power for their own use than for sale, given the relatively low prices utilities are willing to pay. The impact of this clarifying action could be expressed as an incremental fraction of the cogeneration development expected as a result of government action to require equitable sale prices. Some or all potential cogenerators might perceive the clarification of jurisdiction as extending regulatory authority. It is believed, however, that clarification will focus on limiting regulatory jurisdiction over power sales.

The cumulative cost of this action could be well over $1 million from 1978 through 1985, depending on the extent of federal involvement in potential litigation to clarify power sales and equitable prices.

Included in the far-reaching impacts associated with the clarification of power sales is the opportunity for utility customers to receive lower cost electric energy.

Intervention in State Regulatory Proceedings: The Department of Energy (DOE) would intervene as a party to state regulatory hearings that are relevant to industrial cogeneration. In some cases, the department would express support (or opposition to) a particular state action. However, in most cases, DOE would simply present its conclusions regarding the impact of any state action on cogeneration development and energy consumption patterns. Much of the analysis required for DOE testimony would be carried out under the related action, Technical Assistance and Policy Studies for state PUCs.

This action would encourage early state action to promote cogeneration. While the impact is indirect, it is expected to be substantial relative to its cost because of the sensitivity of cogeneration development to the variables over which the state has primary control.

The cumulative cost of this action would be approximately $1 million from 1978 through 1985, for three professionals to prepare and present the DOE testimony. Far-reaching impacts associated with some state actions (e.g., flat rates) would have beneficial effects on industrial cogeneration. The DOE intervention would highlight any adverse impacts.

Technical Assistance and Policy Studies for State PUCs: The Department of Energy would provide several types of assistance to state agencies regulating electric utilities, including technical assistance and policy analysis support.

This action is intended to expedite consideration of electric rates or other policies designed to encourage industrial cogeneration (e.g., attractive pricing of standby arrangements) and to facilitate consideration of the effects, if any, of other electric rates, existing and proposed, on industrial cogeneration (e.g., peak pricing systems).

DOE would assist in analyzing the impact of a state policy on cogeneration. DOE/FERC experience at the national level and with specific potential rate designs would facilitate this analysis. However, much of the policy analysis support would be provided through grants to state-directed projects; state proposals indicate capabilities for conducting analyses.

While the impact of this action would be indirect, it could be substantial, because the states have the greatest leverage over several important variables.

The financial attractiveness of industrial cogeneration is very sensitive to the electric rate for utility power, the charges for utility standby capacity, the charges for wheeling cogenerated power over the utility system to third parties, and the utility price for power sold by industrial cogeneration plants. Also, management assessment of the risks of cogeneration depends on perceptions of utility and commission attitudes. Finally, the extent of industrial cogeneration development in many situations is contingent on utility participation, through ownership, management, or willingness to make transactions. All of these important variables are primarily under the control of the states, not the federal government.

This action would require total federal expenditures of approximately $18 million from 1978 through 1985; $4 million for technical assistance (a staff of 12 professionals); $12 million in grant to states for analysis of the impacts of regulatory actions; and approximately $2 million for a staff to undertake policy analysis at the national level.

This action would lead to state actions which better reflect the states' best interests. Furthermore, this assistance may help identify steps the states can take to mitigate the trade-offs between cogeneration development and any adverse impacts.

Federal Information Clearinghouse: The Department of Energy, with the assistance of the Department of Commerce, would establish and operate a clearinghouse for industrial cogeneration information. Such information would be designed primarily to enable industrial decision makers to identify cogeneration opportunities and assess the suitability and attractiveness of cogeneration for their applications. This clearinghouse would:

> Collect and maintain a library of cogeneration information, including government and private sector reports, applicable federal and state regulations, lists of equipment manufacturers, and perhaps even sales materials from manufacturing and architect/engineering firms.
>
> Design and operate an outreach program to create more widespread awareness of cogeneration technology and attractiveness, and to actively disseminate cogeneration information.

> Review available information to identify areas where the government could stimulate further industrial or utility consideration of cogeneration by carrying out further analysis.

While this action would not by itself stimulate any significant increase in cogeneration, sufficient benefits can be expected, especially in combination with other government actions, to justify its relatively low cost. These benefits will be of two types. First, the existence of a comprehensive source of information would probably expedite the process of evaluation and decision-making for some of those plants or firms already considering cogeneration and for state regulatory commissions. Second, the clearinghouse's active dissemination program should make engineers and/or managers more aware of the relevance of cogeneration to their plants. While it was found that virtually all large firms were well aware of cogeneration, there may be some smaller firms which have less knowledge and which have limited funds for detailed feasibility studies.

Overall, the clearinghouse would have the indirect effect of assisting private sector decision makers in responding to other policies, which have a more direct impact on cogeneration development.

Public sector expenditures for this action will be relatively small and would consist primarily of personnel costs.

Antitrust Exemption for Joint Ventures: The Department of Justice would first study the potential effects of joint cogeneration ventures on competition and market structure. It would then issue regulations or guidelines clarifying any conditions under which such joint ventures would subject any of the participants to antitrust complaints. If, in the judgment of the Departments of Energy and Justice, these guidelines constitute a substantial impediment to cogeneration development, and could be altered so as to encourage more cogeneration without unacceptable competitive effects, then the departments would submit the appropriate legislation to Congress. Such legislation would provide exemptions from antitrust laws, under specified circumstances for cogenerated joint ventures.

The impact of this action would depend on several factors. The greatest of these is the financial attractiveness of cogeneration joint ventures. While industrial managers did not indicate a serious concern over antitrust constraints, this may have been due to the limited interest in joint ventures in the first place. If joint ventures become more attractive, because of enhanced cogeneration economic in general or due to the exploitation of most remaining "own use" cogeneration opportunities, then managers may reconsider antitrust as well as other risks associated with joint ventures.

Some positive impact on cogeneration could be expected, primarily beyond 1985, if such an exemption were granted to cogeneration joint ventures in the near future. The near-term effectiveness of this action and others that remove factors inhibiting joint ventures is compromised by the length of time required to formulate an acceptable arrangement among all parties and to construct the appropriate cogeneration plant, the transmission and distribution systems, and the ancillary equipment.

Another factor is the technical potential for increases in cogeneration development because of joint ventures. Much of the process steam that was removed from consideration for capital investments because of technical considerations can now be considered for cogeneration through joint ventures.

The cost to the government would not appear as a direct increase in expenditures. Rather, the Departments of Energy and Justice will experience the limited opportunity costs associated with assigning existing staff to carry out the review of competitive implications and the drafting of regulations and legislations. Any potential anticompetitive impacts would have to be thoroughly studied before any exemptions were granted from existing antitrust law.

Exemption from Mandatory Coal Conversion Program: The Department of Energy would draft legislation to amend whatever laws are in effect to exempt from coal conversion requirements any boilers to be incorporated in cogeneration systems. If new legislation is not needed, these exemptions would be written into the DOE program regulations. To receive an exemption, a firm would have to apply in writing with a statement of intent to build a cogeneration plant, and subsequently document that the plant was, in fact, built.

There are two potential types of impacts of such an exemption policy on cogeneration development:

1) Some of the cogeneration development would switch from coal to oil fuel. This would represent a reduction in net savings of oil.

2) Cogeneration development would increase since some of those plants subject to coal conversion orders in the absence of this policy would switch from a rejection to an acceptance of cogeneration investments.

The extent of these impacts would depend on the management ROI requirements associated with oil and coal as alternate fuels, and on the ROIs of the coal and oil cogeneration alternatives. Generally, the ROI requirements of coal-fired systems would be high due to the price, availability, and environmental uncertainties. Coal-fired cogeneration plants have substantially greater incremental capital costs than do new, low-pressure process steam coal-fired boilers and oil-fired cogeneration plants. In fact, an oil-fired cogeneration plant could actually have a lower capital cost than a new, process steam coal-fired boiler with a scrubber. With respect to operating costs, while coal is the cheaper fuel, a coal system also has significant other operating costs such as for coal handling and emission control.

If the financial attractiveness of the coal-fired cogeneration plant (over a new low-pressure process steam boiler fired by coal) is matched by that of the oil-fired cogeneration plant (over the same management alternative), then much of the cogeneration development expected in mandatory coal conversion configurations without government action will switch to oil. While this switch is not desirable from a national policy point of view, the overall development of cogeneration will increase to the extent that the ROIs of oil-fired cogeneration plants (with the exception of coal conversion) exceed the mandatory coal conversion ROI.

The cost to the government would appear as lost revenue through the investment tax credit. Obviously, any additional cogeneration development would produce additional income tax revenue during the operating life of the separate plants. The Department of Energy would be required to allocate staff to process and monitor the candidates for exemption from coal conversion. The exemption from coal conversion and the additional fuel requirements of cogeneration plants could produce an undesirable consumption of oil and natural gas.

Support for Demonstration Projects and Technical Assistance: Possible energy savings to the nation through the near-term implementation of government programs designed to facilitate cogeneration technologies and systems could be substantial beyond 1985.

The greatest potential for government stimulation of industrial cogeneration is the removal of the technical constraints inhibiting development. As indicated previously, approximately 3,500 trillion Btu of process steam by 1985 will be technically suitable for cogeneration development. This value compares to the approximately 2,300 trillion Btu of process steam in 1985, which is technically suitable but not financially acceptable without government action. Appropriate DOE programs designed to remove or reduce many of the technical barriers of cogeneration could increase the amount of process steam suitable for capital investment considerations.

DOE should promote a competitive suppliers' market of systems and equipment capable of meeting technical and economic performance criteria required by potential cogeneration candidates. For instance, these four areas technically inhibit cogeneration development:

> At some manufacturing facilities the comparative economic and technical performance of mechanical drive is superior to electric motors. The operating environment, the flexibility of operation, and the available space, in some instances, dictate the present requirement for steam drive. In environments of high temperature and atmospheric particulates, considerable protection must be given electric motors, whereas mechanical drive is unaffected. Variable high-speed electric motors are considerably more expensive than comparable mechanical drive. For requirements of 1,000 horsepower and greater, electric motors generally require considerably more space than comparable mechanical drive. Efforts to improve the competitiveness of electrical motors can promote the development of cogeneration.
>
> In manufacturing plants where the electricity and steam required for process fluctuate out of phase, supplemental systems are currently employed. The commercialization of energy storage systems capable of leveling the loads on cogeneration plants eliminates the need to purchase supplemental utility-generated power (perhaps during utility peak periods), passing steam energy through condensors when not required in the manufacturing process, and firing auxiliary process steam boilers when the electric demand is low.

At manufacturing locations where waste heat can be recovered to produce steam of 400 psig and less, this steam cannot, in practice, be used to produce electric energy prior to use in process. The stimulation of technologies capable of utilizing low-grade steam to produce electric power in an efficient and economic manner could enhance the development of cogeneration.

Manufacturing facilities with low steam requirements, i.e., under 50,000 pounds per hour, are generally not adaptable to steam turbine cogeneration systems employing standardized equipment. Prior to the oil embargo in 1973, gas turbines and diesels with waste heat recovery were widely acceptable by small users as an alternative to purchasing electric energy. The price and availability of fuel has affected the relative economic performance of these systems. The development of more adaptive systems and the flexibility to use more secure and cheaper fuels could enhance the development of cogeneration.

At manufacturing facilities where residual fuels are employed for steam generation, the efficiencies are below those resulting from primary fossil fuels (e.g., about 60% for waste fuels used in the paper industry versus 80% for oil). Although actions to improve boiler efficiencies do not increase the amount of process steam technically suitable for cogeneration, they would provide the mechanism to save energy.

This list is not intended to be complete, but it does illustrate the potential for cogeneration development through DOE-sponsored demonstration programs and technical assistance. For instance, the complete removal of load fluctuations as a technical barrier to cogeneration development would increase the technically suitable steam by 11% in 1985, representing about 620 MW of cogeneration capacity and a corresponding annual energy savings of 20 trillion Btu. However, the removal of this inhibiting factor is not anticipated by 1985.

DOE already possesses the mechanisms necessary to undertake demonstration programs and provide technical assistance to augment these ongoing efforts. Significant expenditures will be necessary to determine the applicability of cogeneration systems in industry beyond 1985, to ascertain the additional requirements for successful implementation, and to promote a viable supplier market.

Significant expenditures will be necessary to determine the applicability of cogeneration systems in industry beyond 1985, to ascertain the additional requirements for successful implementation, and to promote a viable supplier market.

Operator Training Grants: The Departments of Energy and Commerce would jointly establish a program under which grants would be made directly to firms with insufficient personnel to operate and maintain cogeneration plants built after April 20, 1977. DOE would specify a set of academic requirements for training operators for cogeneration plants and make grants for such training.

Since only about 1 to 2% of the market that finds cogeneration technically suitable and financially attractive expects difficulty in locating skilled personnel, the impact of such grants on cogeneration would be small, and its cost would depend on the number of operators for which firms request training grants. The maximum cost would be $2 million, if virtually all operators of cogeneration plants built from 1978 through 1985 received $2,000 worth of training under this program.

POTENTIAL GOVERNMENT PROGRAMS

So far the effect of individual government actions has been considered. Government programs comprising several individual actions could stimulate industrial cogeneration development prior to 1985 and establish the foundation for greater development beyond 1985. The six programs examined include a range of financial and regulatory incentives. Individually, they are estimated to produce energy savings ranging from 60 to 200 trillion Btu. The incremental cogeneration capacity which would develop as a result of the six programs ranges from 2,000 MW to 6,000 MW and is shown in Table 3.20.

These programs would stimulate development of process steam by 1985 in amounts ranging from 300 trillion Btu, with the minimum development program, to 900 trillion Btu, with the maximum development program (Table 3.21). Under any potential program, the pulp and paper industry would develop the greatest amount of cogeneration ca-

TABLE 3.20: INCREMENTAL COGENERATION CAPACITY: DEVELOPMENT BY 1985 WITH GOVERNMENT PROGRAMS (capacity in megawatts)

	Program					
Government Actions	**1**	**2**	**3**	**4**	**5**	**6**
30% ITC (20% additional ITC) for all types of cogeneration plants	2,578	2,578	2,578	2,578		
Accelerated depreciation schedule of 12 years for plant and equipment	1,052	1,052				
Government loan guarantees	617	617			617	617
50% ITC (40% additional ITC) for coal scrubbing equipment	0					
Clarification of FERC and state PUC jurisdiction over industrial cogeneration selling power		medium	medium	medium	medium	medium
Exemption from FERC and state PUC jurisdiction for industrial cogeneration		215	215	215	215	215
Equitable transaction prices for purchase of power from industrial cogeneration plants		407	407	407	407	407
Establishment of flat rate structure						1,808
Peak and off-peak rate structures					low	low
Policy studies and technical assistance to states for rate setting		medium	medium	medium	medium	medium
Intervention in state regulatory hearings					low	low
Tax on industrial oil and gas consumption, with exemption for cogeneration		461	461	461	461	461
Cogeneration plants exempt from coal-conversion requirements					high	high
SO_2 emission standards relaxed for new plants				752		
Antitrust exemption for joint ventures					medium	medium
High priority for fuel allocation to industrial cogeneration plants		136			136	136
Demonstration support for available technology (i.e., small generators, distillate-fired turbines, coal handling, coal SO_2 scrubbers, sludge and ash handling, fluidized-bed boilers, energy storage, and load fluctuation responsiveness of cogeneration plants)	high	high	high	high		
Federal information clearinghouse	low	low	low	low		
Operator training (grants)	low	low				
Total	**4,247**	**5,466**	**3,661**	**4,413**	**1,836**	**3,644**

TABLE 3.21: RANGE OF ADDITIONAL COGENERATION BY 1985 THROUGH GOVERNMENTAL PROGRAMS

Industry	Minimum Development Program (No. 5: Heavy Regulatory)			Maximum Development Program (No. 2: Heavy Financial)		
	Process Steam Cogenerated (trillion Btu)	Electric Energy Cogenerated (billion kWh)	Cogeneration Capacity (MW)	Process Steam Cogenerated (trillion Btu)	Electric Energy Cogenerated (billion kWh)	Cogeneration Capacity (MW)
Food	27	1.18	195	58	2.56	420
Textile	3	0.18	29	5	0.30	47
Pulp & Paper	112	6.01	763	328	16.12	2,046
Chemical	61	2.81	357	225	10.34	1,313
Petroleum Refining	49	2.71	346	140	8.08	1,028
Steel	27	1.14	146	103	4.82	612
Total	**279**	**14.03**	**1,836**	**859**	**41.22**	**5,466**

pacity and the textile industry the least (Table 3.22). This result is attributed to the relatively low, risk-adjusted ROI hurdle rate for cogeneration projects within the pulp and paper industry compared to the high ROI required in the textile industry. In general, however, relatively little cogeneration capacity would result from stimulation by government programs.

TABLE 3.22: INCREMENTAL COGENERATION CAPACITY BY 1985 WITH GOVERNMENT ACTION
(capacity in megawatts)

	Program					
Industry	**1**	**2**	**3**	**4**	**5**	**6**
Food	310	420	248	271	195	298
Textiles	26	47	35	41	29	44
Pulp & Paper	1,616	2,046	1,275	1,585	763	1,391
Chemical	996	1,313	956	1,123	357	846
Petroleum Refining	823	1,028	715	863	346	689
Steel	466	612	432	530	146	376
Total	**4,237**	**5,466**	**3,661**	**4,413**	**1,836**	**3,644**

To determine the appropriateness of federal government investment to stimulate industrial cogeneration development, potential programs must be assessed relative to broad government concerns. Table 3.23 presents some criteria for such assessment and briefly explains what each criterion measures.

The overall conclusion was that Program 1, consisting of only heavy financial incentives, appears to have the greatest advantages. Its net annual energy savings of over 140 trillion Btu by 1985 are second only to Program 2. Program 1, however, has the greatest likelihood of successful stimulation of cogeneration because it does not have the complexities and uncertainties surrounding the programs containing regulatory actions.

In general, Programs 1 and 2, consisting of heavy financial incentives, appear to have the greatest overall advantages, whereas Programs 3 through 6, with regulatory emphasis, appear less desirable. The relative advantages and disadvantages of each program are discussed below. Table 3.20 should be used as a reference throughout.

TABLE 3.23: BROAD GOVERNMENT PROGRAM EDUCATION CRITERIA

Criteria	Measure
Extent of federal cost over time.	Reflects financial costs to government, the duration of financial commitments and their predictability and controllability.
Degree to which a government program will increase industry participation and promote competitiveness in cogeneration equipment supply.	Reflects extent to which different companies are encouraged to form a competitive market supplying cogeneration equipment and undertaking related R&D.
Certainty that a government role will increase prospects for industrial cogeneration development.	Reflects probable effectiveness of government program. Key considerations include the relationship of the program to the major problems in industrial cogeneration development.
Administrative feasibility of a government program.	Reflects required personnel and costs, need for new organizational units and their duration, minimization of controversial regulatory procedures.
Extent of government involvement in management of operations in the industry.	Reflects degree of government involvement in operating decisions, regulatory requirements, technical requirements; extent of government dominance of the market in a manner that creates industrial dependency on government funding and guidance.
Existence of authority to implement the program.	Reflects the current existence of authority to implement the program.
Degree to which a government program can produce national benefits by focusing on industry problems.	Reflects the amount and type of energy savings that can be achieved through stimulated industrial cogeneration development.
Extent of impacts that are produced through a government program.	Reflects the degree to which potential detrimental local and national impacts are incurred.

Programs 1 and 2—Heavy Financial Incentives

Programs 1 and 2 contain many of the same actions, except that Program 2 has additional actions directed toward removing some of the risk-related issues associated with industrial cogeneration. Despite the expected significant costs and the required legislation, Programs 1 and 2 are desirable in terms of their relative certainty of success, their focus, their administrative feasibility, and their expected large energy savings. Further, the long-term expected benefits through the stimulation of an active and competitive supplier market, properly focused on the requirements of industrial cogeneration candidates, enhance the viability of these programs.

The additional regulatory actions of Program 2 increase the expected energy savings by focusing on the risk-related issues inhibiting cogeneration development. However, the additional financial and administrative costs to the government, the required federal legislation and regulations, and the increased possibility of federal government involvement in industrial operations and state regulation reduce the program's overall attractiveness.

Programs 3 and 4—Moderate Financial and Regulatory Incentives

Programs 3 and 4 contain the same federal actions except that Program 4 includes an environmental action: reduction of the further emission standards required for cogeneration plants. The major regulatory actions included in these programs focus on issues surrounding the possible transactions between industrial cogeneration plants and utilities. In addition, an action is included which exempts industrial cogeneration plants from taxes on oil and gas use.

Because of the relative uncertainty for success, the great complexity associated with the administration of the programs, and the potential involvement in industrial operation and state regulation, these programs fall below the expected overall advantages of Program 1. Further, the addition of the relaxed sulfur emissons standards for industrial cogeneration plants increases the potential for severe local environmental and socioeconomic impacts; this also reduces the acceptability of Program 4 compared to Program 3.

Programs 5 and 6—Heavy Regulatory Incentives

Programs 5 and 6 include the federal government loan guarantee, as a minimum financial incentive, and all the regulatory actions contained in Program 3. In addition, Program 6 contains the government action that establishes flat rates for purchased electricity.

The high administrative costs, the uncertainties surrounding the success of the actions, the federal involvement in industrial operations and state regulation, the required federal legislation and regulations, the lack of emphasis on stimulating supplier markets, and the moderate to low expected energy savings decrease the acceptability of Programs 5 and 6. Out of the six programs considered here, these two appear to be the least desirable, despite the fact that they are the lowest cost programs to the federal government.

Other Program Considerations

Although the potential government programs addressed here consist of discrete actions, the individual actions are not entirely independent of one another in the context of a program. Consideration must be given to the overlapping and contradicting effects of the combined actions.

For instance, Program 6 contains both time-of-use rates and flat rates as separate actions. Taken alone, flat rates encourage the industrial cogeneration of electricity. When these actions are combined and implemented, their impact on cogeneration development could be different. In fact, time-of-use flat rates probably would have no significant impact on the cost of purchased electricity, and hence on the development of industrial cogeneration.

Further, the combined effect of the federal actions to clarify FERC and PUC jurisdiction over the sale of electric power by industrial cogeneration plants and to establish equitable transaction prices could be greater than the effect of the two actions taken in isolation. These actions can lower the risks and uncertainties associated with the sale of electric power and can enhance the ROI of cogeneration development.

In a general sense, the expected benefits from any set of proposed government actions are not strictly additive. With the exception of Program 6, which contains the flat rate action, analysis indicates that the combined benefits are between 5 and 15% above the sums indicated in Figure 3.12. This difference is caused by overstating the impact of more than one action on the same cogeneration candidate, or understating the impact of multiple actions on candidates where no single action is sufficient to stimulate cogeneration investment.

UTILITY INVOLVEMENT IN COGENERATION

Although the question of ownership or operation of cogeneration facilities by electric utilities was not examined in this study, it was raised repeatedly by representatives of the utilities interviewed and of the six industries examined. This section briefly examines the disadvantages of industrial cogeneration as viewed by both industry and utility management. The advantages of utility involvement in cogeneration—for example, joint, third-party ownership—are then explored. The issue is complex and requires further examination before analysis of cogeneration potential can be considered exhaustive.

This report examined several factors that made cogeneration unattractive for internal power requirements only. These factors were associated with the low returns and the risk of industrial cogeneration. Large capital requirements, lack of economics of scale, high fuel costs, environmental constraints, and practical limitations associated with cogeneration technologies for small users cause low returns. While larger scale facilities alleviate some of these negative aspects, the prospect of producing and supplying excess electric power to the utility system creates a fear of regulation and accentuates concern about entering a different business.

The utilities are also concerned about industrial cogeneration, particularly about losing those industrial customers who build cogeneration plants, and the interconnections of industrial cogeneration plants with the utility grid.

Significant industrial cogeneration development in a utility service area could severely affect the financial position of the utility. The loss of baseload customers could alter the utility electric load patterns, so that the cost to the utility to produce one kW hour of electric energy would increase. This greater cost is incurred because a greater portion of the remaining customers would be supplied by more expensive generating plants, i.e., peaking and intermediate capacity. The revenue lost by the utility as a result of the loss of large customers coupled with the increased cost per kW hour of electric energy produced could sharply reduce the financial return to the utility.

However, negative effects on the utilities as the result of development of industry-owned and operated cogeneration plants may be only short term. Improved load management practices and general load growth could, in the long term, offset utility cost increases caused by the loss of baseload customers. Furthermore, significant levels of industry-owned cogeneration plants could alleviate many of the capital constraints currently incurred by utilities. That is, the installation of large amounts of industry-owned cogeneration capacity could reduce the utilities' plans for new, expensive, baseload generating capacity, primarily coal-fired and nuclear power plants.

Interconnections between utility grids and industrial cogeneration plants also create concerns for utility managers. Generally, the utilities want to maintain dispatching control over the electric power entering the grid to ensure system stability and security. If the purchased power is significant in amount, the utilities want guarantees as to the time of delivery, the amount, and the length of the agreement, among other considerations. In addition, virtually all utilities are concerned about precipitous government action that recognizes only the industrial point of view, and fails to take into account the total economics of the utilities and industrial companies, and the impact on the various consumers of electric power.

Most industrial and utility concerns would be eliminated, or greatly negated, under utility ownership, joint ownership, or a third-party ownership arrangement. From the industrial viewpoint, the fear of regulation of the manufacturing company would be eliminated. Also, certain cogeneration investments would be economic for a regulated utility (at a 7 to 10% return) where they would not be economic for the industrial company. Further, when several industrial users are situated within a few miles of each other, greater economies of scale could be provided to multiple users from a larger plant. While a multiple user plant could solve the low economies of scale problem without utility involvement, the practical needs to arrange appropriate rates for all partners in the plant would be greatly facilitated if the utility was at least a joint owner, if not the sole owner. Finally, capital constraints would be greatly mitigated under joint or third-party ownership.

From the utility viewpoint, most of the concerns about industrial ownership would be alleviated if the utilities owned or participated in the operation of the cogeneration plants. Utilities with ample capacity (high reserve margins), however, might still regard industrial cogeneration negatively unless the return from these plants was increased.

One factor could be perceived as increasing returns. In some states, a utility-owned plant that supplies power only to the on-site customers is exempt from regulation; this exemption allows the utility to receive a greater than regulated rate of return, if the economics otherwise allow it. Therefore, depending on the cost of excess power relative to other sources of power, the utility may have no incentive to build a plant that supplies excess power. This would be most true for utilities in states with a large reserve margin, and absence of a need for additional baseload capacity for more than 10 years.

Even without a careful analysis, there is one government action that could change the existing economic incentives markedly from the utility standpoint and that is exemption of *all* cogenerators from federal or state regulation. Indeed, there would then be the highest possible incentive on the part of the utility to produce as much electric power as possible from the lower cost cogeneration plants, since any other new (and regulated) baseload plants would almost always have higher cost electric power for the consumer at a lower return for the utility.

The issue of utility involvement as a means of encouraging cogeneration development is worth pursuing in light of the significant amounts of steam technically suitable for cogeneration. Figures 3.13 to 3.18 inclusive illustrate the potential for cogeneration in each of the six industries examined. Industry is expected to cogenerate only about 35% of the technically suitable steam; the white area on each 1985 bar graph shows the remaining potential. In total, about 40% of the technical potential is not economic for industry alone—even with government actions—but could be with utility involvement (i.e., ownership or operation).

FIGURE 3.13: PROCESS STEAM FROM INDUSTRIAL COGENERATION IN THE FOOD INDUSTRY IN 1976 AND 1985 (Btu $\times 10^{12}$)

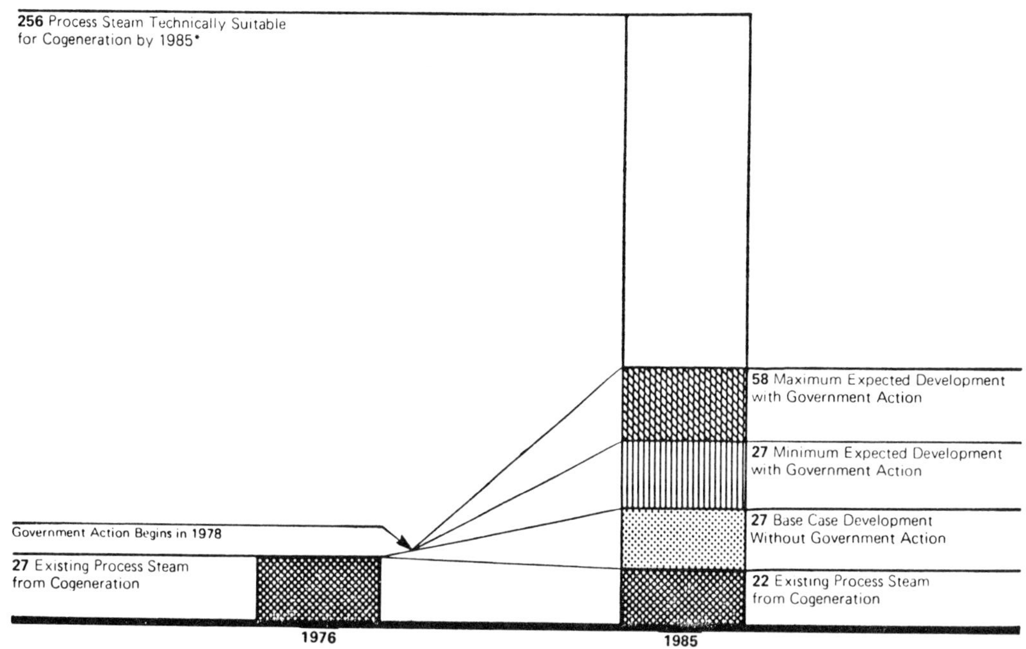

*Includes 20 trillion Btu of process steam projected to be produced in 1985 by cogeneration plants operating in 1976.

FIGURE 3.14: PROCESS STEAM FROM INDUSTRIAL COGENERATION IN THE TEXTILE INDUSTRY IN 1976 AND 1985 (Btu × 10^{12})

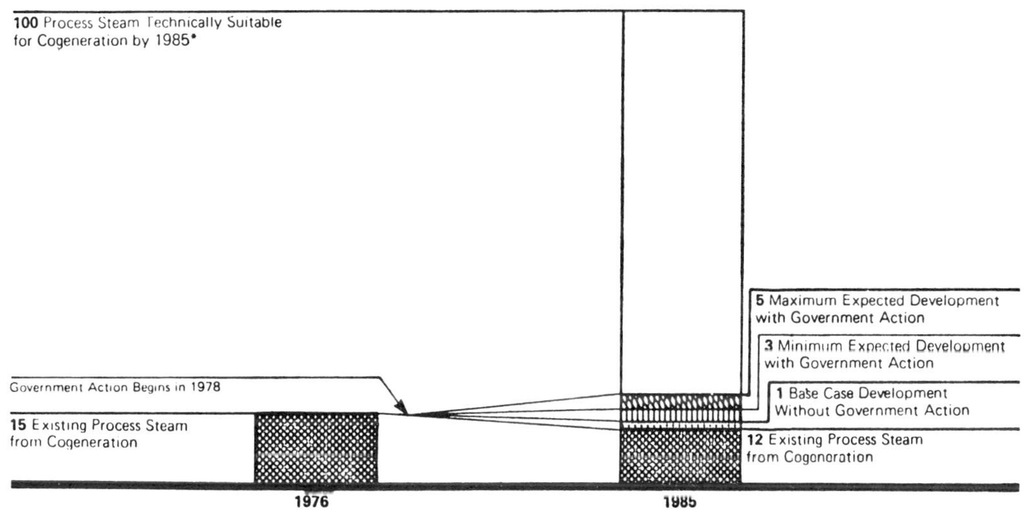

*Includes 12 trillion Btu of process steam projected to be produced in 1985 by cogeneration plants operating in 1976.

FIGURE 3.15: RANGE OF PROCESS STEAM FROM INDUSTRIAL COGENERATION IN THE PAPER INDUSTRY IN 1976 AND 1985 (Btu × 10^{12})

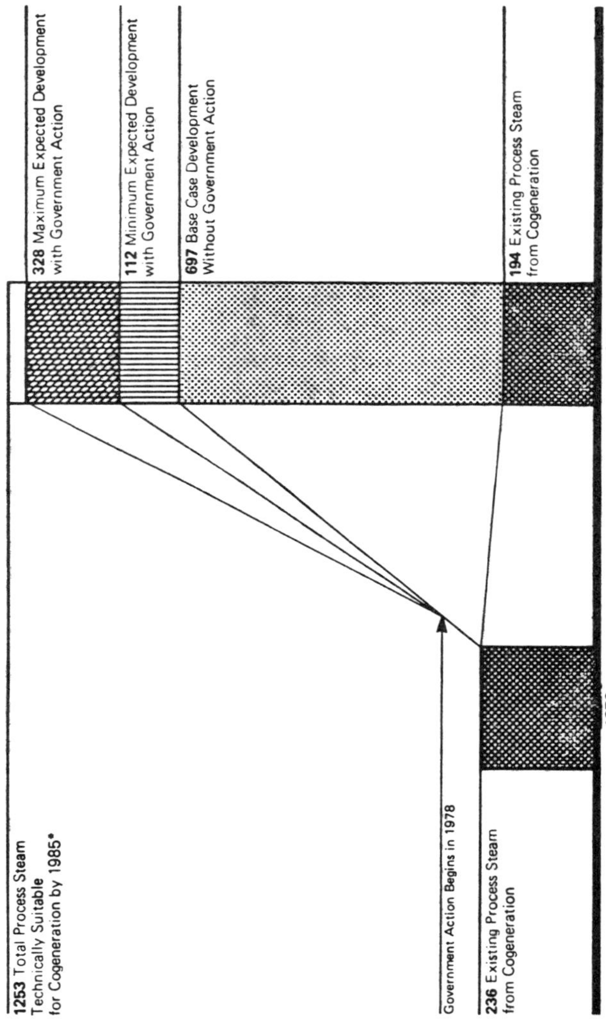

*Includes 194 trillion Btu of process steam projected to be produced in 1985 by cogeneration plants operating in 1976.

FIGURE 3.16: PROCESS STEAM FROM INDUSTRIAL COGENERATION IN THE CHEMICAL INDUSTRY IN 1976 AND 1985 (Btu × 10^{12})

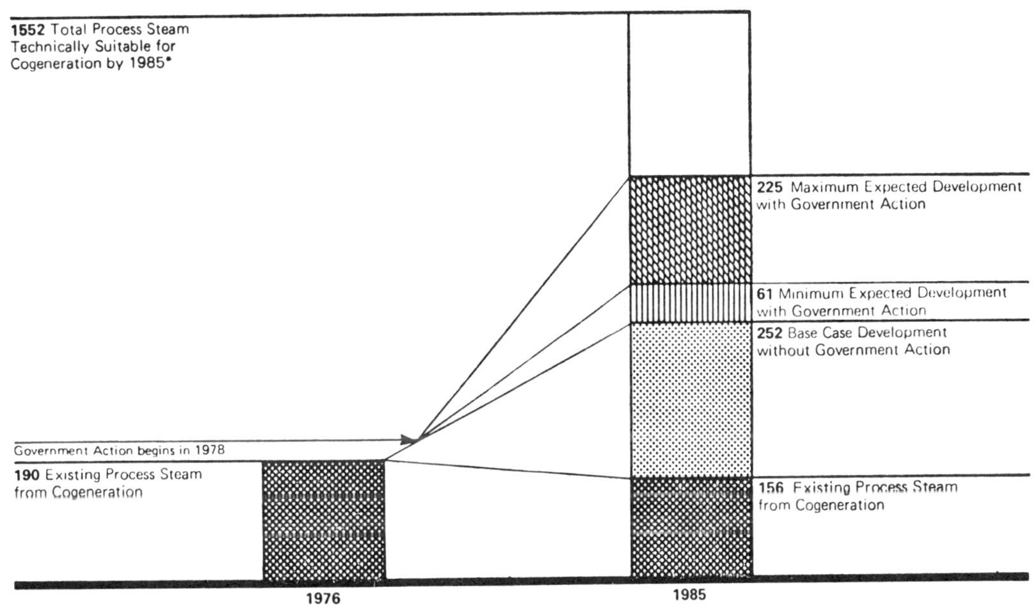

*Includes 156 trillion Btu of process steam projected to be produced in 1985 by cogeneration plants operating in 1976.

FIGURE 3.17: PROCESS STEAM FROM INDUSTRIAL COGENERATION IN THE PETROLEUM REFINING INDUSTRY IN 1976 AND 1985 (Btu × 10^{12})

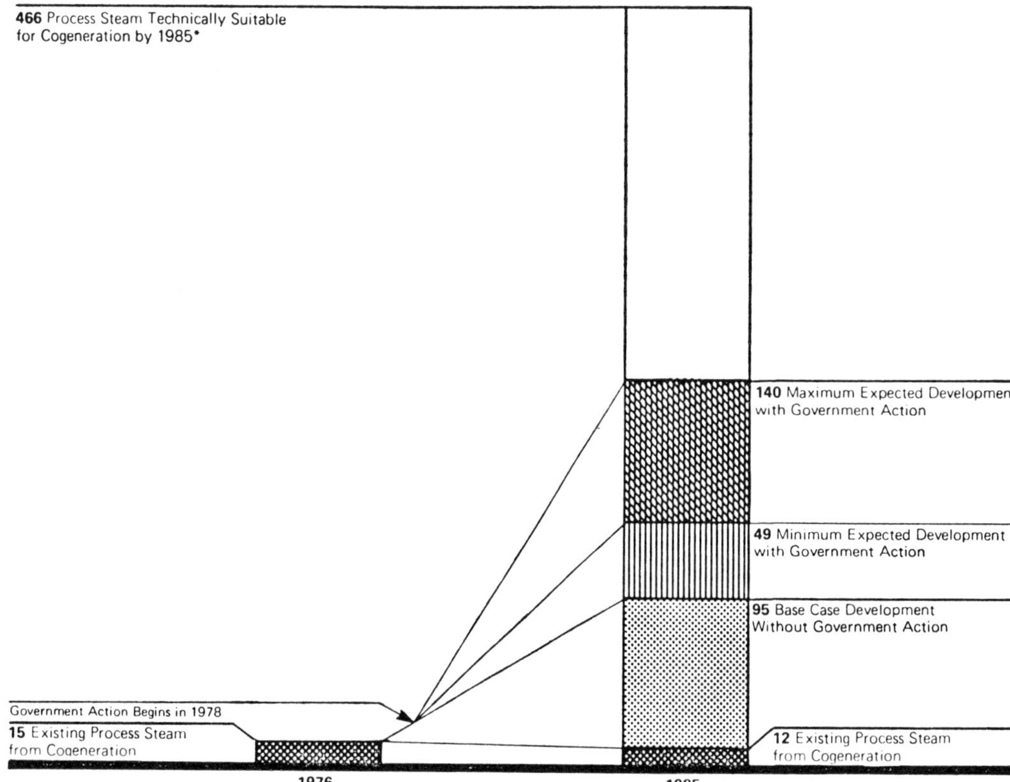

*Includes 12 trillion Btu of process steam projected to be produced in 1985 by cogeneration plants operating in 1976.

FIGURE 3.18: PROCESS STEAM FROM INDUSTRIAL COGENERATION IN THE STEEL INDUSTRY IN 1976 AND 1985 (Btu x 10^{12})

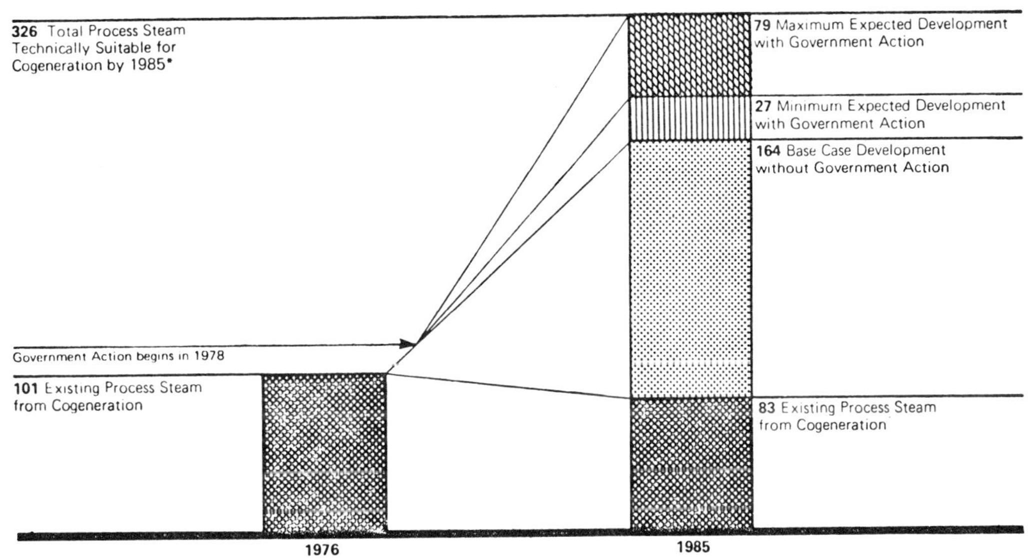

*Includes 83 trillion Btu of process steam projected to be produced in 1985 by cogeneration plants operating in 1976.

A Technical Overview of Cogeneration

Resource Planning Associates Study

The material in this chapter was taken from the report by Resource Planning Associates, Inc. prepared for the Department of Energy, Division of Industrial Energy Conservation, entitled *A Technical Overview of Cogeneration: The Hardware, the Industries, the Potential Development,* and dated December 1977. Project Director for RPA was Howard Hagler, and Project Manager was Howard Lowitt. The introduction and summary is a direct quote from the report. The rest of the chapter has been edited and summarized somewhat.

INTRODUCTION AND SUMMARY

Cogeneration, the simultaneous production of more than one form of useful energy, has long been used by industry as a means of producing both electric and thermal energy to meet on-site process needs. Also known as total energy systems, integrated energy systems, dual-purpose power plants, and waste-heat utilization systems, cogeneration systems provided over 15% of the nation's electric energy in 1950; in recent years, however, that fraction has declined gradually to a current 5%.

The Department of Energy (DOE) has initiated several programs aimed at promoting energy conservation in industry. Because the by-product heat from a power-conversion process is captured for productive use in a cogeneration system, instead of exhausted to the environment as it is in a conventional power plant, cogeneration represents an important energy-conservation technique. By cogenerating, an industrial plant can save the fuel that would have been needed to produce the amount of heat captured.

Recognizing the significant energy-savings potential offered by cogeneration, DOE has undertaken a major research, development, and demonstration (RD&D) program to investigate and promote cogeneration in industry. Resource Planning Associates, Inc. (RPA), has been working with DOE to accomplish four of the program's major objectives:

- Survey current, near state-of-the-art, and future cogeneration equipment, and identify any gaps or deficiencies.
- Characterize the energy requirements of the manufacturing sectors of five of the country's most energy-intensive industries (specifically, chemical, petroleum refining, paper and pulp, textiles and food and kindred products).
- Identify principal targets for, and barriers, to the increased market development of cogeneration systems.
- Estimate the potential maximum and the probable energy savings that could be achieved in the five selected industries through cogeneration.

In performing this study, we conducted an extensive literature search; interviewed hardware manufacturers and industrial-plant owners and operators; and visited industrial plants that are currently using cogeneration systems. In addition, we assessed the effects of economic and institutional barriers, as well as potential government actions, on the probable development of cogeneration, relying heavily on the findings of RPA's recent study of the potential for cogeneration development in six major industries, conducted for the Federal Energy Administration [see Chapter 3].

Based on our research and analysis, we have estimated that the widespread application of cogeneration technology in the five industries studied could result in a maximum potential savings of 2.4 million barrels of oil equivalent per day (or a maximum incremental capacity of 140,000 MWe) by 1985. To estimate the probable level of cogeneration development, we incorporated economic, institutional, regulatory, and technical considerations into our analysis.

With a moderate degree of government incentives, defined as: 30% investment tax credit (ITC), clarification of Federal Power Commission (FPC)/Public Utility Commission (PUC) jurisdiction for nonexporters, and an equitable price for export to the grid, and varying technology mixes (see Figure 4.1) probable cogeneration development can range from 9,500 to 20,000 MWe in 1985. These incremental capacities translate into probable energy savings ranging from 0.20 to 0.60 million barrels of oil equivalent per day in 1985.

In investigating cogeneration hardware, we concentrated on three specific technologies: steam turbines, gas turbines, and diesel engines. Of these three technologies, the steam turbine is currently the most widely used in industry. Although the gas turbine and diesel engine offer even greater energy-savings potential, as a result of their high electric-to-thermal ratios, their applications have to date been limited.

One of the major barriers to the widespread application of gas turbines and diesel engines is that they generally require premium fuels. Although some gas turbines have dual-fuel capabilities (i.e., can burn either natural gas or pertroleum distillates), they are unlikely to gain wide market acceptance until they are able to burn coal-derived gases or liquids and residual oil. Currently, gas turbines are used primarily in the petroleum and chemical industries. Diesel engines are not widely used by any of the industrial sectors studied, and are not likely to gain market acceptance until they are able to burn liquefied, gasified, solvent-refined, or pulverized coal. Research is presently under way to develop gas turbines and diesel engines that have these capabilities.

In the industrial sectors studied, the back-pressure steam turbine (or extraction noncondensing turbine) matches thermal and electrical demand most accurately and is therefore the most widely used type of cogeneration hardware. One factor, however, limits the applicability of the steam turbine for small users—it is not economically competitive with the gas turbine or diesel engine in the 5 MW-or-less power range.

FIGURE 4.1: COGENERATION SAVINGS COMPARED TO TECHNOLOGY MIX (1985)

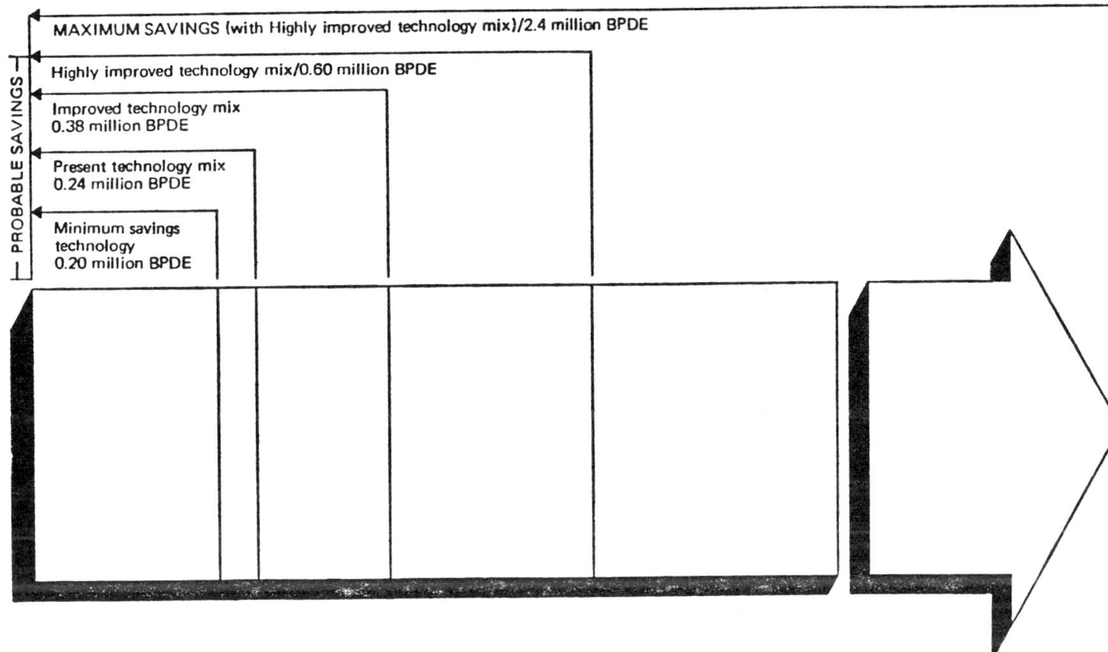

TECHNOLOGY MIX	TOPPING OF PROCESS STEAM	TOPPING OF PROCESS HEAT	BOTTOMING OF WASTE HEAT
Minimum saving	100 percent steam turbine	100 percent gas turbine	100 percent steam Rankine
Present	90 percent steam turbine 10 percent gas turbine	100 percent gas turbine	100 percent steam Rankine
Improved	75 percent steam turbine 15 percent gas turbine 10 percent diesel	50 percent gas turbine 50 percent diesel	50 percent steam Rankine 50 percent ORC toluene
Highly improved	50 percent steam turbine 50 percent diesel	50 percent gas turbine 50 percent diesel	50 percent steam Rankine 50 percent ORC toluene

General consensus among industry representatives is that federally funded research should focus on developing hardware with multi-fuel capability (including the capacity to burn nondistillate fuels), improving the operation cycle time (i.e., availability and reliability) of hardware, reducing initial capital and maintenance costs, and further developing energy storage techniques to compensate for load fluctuations. Improving the efficiencies of the various technologies seemed to be less important to industry managers. All agreed, however, that it is essential that industry be kept informed of technological advances in cogeneration hardware.

Government policy towards coal conversion will significantly affect the development of cogeneration. In general, mandatory conversion to coal is likely to encourage cogeneration by increasing the use of direct coal-fired steam turbines. However, such a policy would not maximize the potential savings available through cogeneration. If exempted from the requirement to convert plants to coal, industry would be able to use oil and gas, thus gas turbines and diesel engines, which offer greater energy savings.

In anticipation of a potential loss of their baseload, utilities are becoming more receptive to cogeneration joint ventures, and are considering both in-plant and centrally located power plant systems. However, several important issues and concerns must be resolved if operating agreements are to be reached and market development is to proceed. These issues relate to:

- Management and operational control of equipment with assurances to industry of a reliable source of thermal energy.
- Delays associated with environmental and siting regulations and procedures.
- Power of state agencies to prohibit the sale of industrially generated excess power.

INDUSTRIAL REQUIREMENTS

This section discusses the potential for cogeneration in the chemical, petroleum refining, paper and pulp, textile, and food and kindred products industries. Information was derived mainly from interviews with managers and visits to plants.

These five industries were chosen for analysis because they were all among the ten largest energy users in the country in 1976 and because they all apparently required enough thermal and electric energy to make cogeneration an economic option. Beyond these similarities, the industries show important variations which gave the opportunity for consideration of different configurations of cogeneration equipment. For example, some of the industries (e.g., petroleum refining) have a wide range of plant sizes, while others (e.g., food and kindred products) have mainly small plants. Some have relatively stable production processes, while others have frequent modifications that increase the opportunities for the introduction of new equipment.

Furthermore, while the chemical and textile industries are geographically concentrated; petroleum is somewhat dispersed; and food, and paper and pulp are widely distributed throughout the country. These differences are also important, because a widely distributed industry can minimize the effects of regional differences in energy policies or prices. Where industrial plants are concentrated in particular areas of the country, the price and availability of fuel, and environmental and regulatory controls can restrict the development of cogeneration equipment.

Investigation of these five industries showed that managers had the following certain common concerns that tempered their enthusiasm for cogeneration:

- Cogeneration technology is not well known.
- Fuel supplies and prices are uncertain.
- Fuel flexibility will be increasingly important, and cogeneration equipment will have to demonstrate this characteristic.
- Steam turbines are currently used in cogeneration systems in all five industries because they match the thermal and electrical demand more accurately than other equipment. However, steam turbines do not always satisfy the electrical demand of most plants and back up electricity must be purchased.
- Gas turbines and diesels, which would satisfy demand for electricity better than steam turbines, are perceived to be expensive to buy and

operate, to require scarce or unavailable fuels, and to have restricted power levels. None of the five industries makes much use of diesels.

- Export of surplus electricity, which might make gas turbines and diesels economically more attractive, is seen to face technical, institutional, and regulatory barriers. In particular, the control of operation (plant manager vs utility manager) is seen as an obstacle.
- Improved energy storage systems, which might also make gas turbines and diesels economically more attractive, are not yet available.
- Energy conservation measures (both for processes and housekeeping), which have recently been instituted in all five industries, may have both positive and negative effects on the market development of cogeneration. Between 1972 and 1976, the net reductions in the use of fuel in these five industries ranged from 9 to 12%. On the one hand, managers who have recently invested in new equipment may not be ready to invest in additional equipment for cogeneration. On the other hand, where modifications have reduced thermal demand, plant load profiles may be better matched to gas turbine or diesel cogeneration systems. Furthermore, in such cases, smaller and cheaper cogeneration equipment may be adequate.
- Cogeneration may require the burning of additional fuel which (while leading to a net reduction in air emissions in the area) would lead to increased air emissions at the plant site. This increase might lead to siting and environmental problems.
- The electric grid may be the most reliable and flexible source of electricity.

All five industries show certain common trends that will affect the market development of cogeneration systems. For example, patterns of fuel consumption changed in a similar manner between 1972 and

1976. According to data from the American Manufacturing Association, the use of natural gas declined considerably and was partially replaced by petroleum, the use of electricity increased, and the use of coal decreased slightly (except in the food industry). Changes will continue over the next few years: natural gas will continue to decline, but coal will almost certainly increase. The future fuel mix is hard to predict and this uncertainty is reflected in management decisions on combustion equipment.

In addition to such long-term uncertainties, all of the industries face short-term uncertainties in production processes and schedules. These fluctuations lead to mismatches between thermal and electrical demand that occur on an hourly, daily, or seasonal basis. These mismatches varied so much within each of the five industries that they must be considered on a plant-specific basis. However, mismatches generally reduce the potential for cogeneration.

Plants presently must compensate for these mismatches by going outside the cogeneration system. They can either add extraction-condensing turbines or process steam boilers; or purchase supplemental electricity from the grid. (The integration of improved energy storage into cogeneration systems would solve some of these problems.) All of these options impose an economic penalty, thus it is important to choose cogeneration hardware that best matches the thermal and electrical load at any plant.

Table 4.1 shows the thermodynamic and electrical profiles of the five industries chosen, expressed first as ranges of actual characteristics and then as the characterizations selected for this study. Figure 4.2 shows the variation in the electric-to-thermal ratio as the demand for electricity changes. For this exhibit, a regression analysis was run on each set of industry data (plant distribution) to obtain a smooth curve; the curve for each industry is analyzed in the sections that follow. Waste heat profiles for each industry are shown in Table 4.2. Waste heat is any thermal energy rejected from a manufacturing process. At temperatures over 250°F, this energy can be captured efficiently by waste recovery equipment.

Chemical Industry

The chemical industry (SIC 28) is very diverse, with over 11,000 plants of 1,300 companies producing over 10,000 products. The chemical industry is divided into eight major classifications as shown in Table 4.3.

TABLE 4.1: INDUSTRIAL CHARACTERIZATIONS

Industry	Firms Contacted	A Average Steam Demand (1,000 lb/hr)	B Average Electric Demand (MW_e)	C Operation Time (hr/wk)	D Process Temperature Conditions (steam) (°F)	E Process Steam Pressure (psig)	F Process Temperature Conditions (heat) (°F)	G E/T* (10^{-2})
Chemical (SIC-28)	25	30–2,500	5–125	168	280–390	50–200	100–600 1,000–2,000	1.5–15
For study		700	40	168	360	150	1,000	10
Petroleum Refining (SIC-2911)	6	250–3,500	10–150	168	280–360	50–150	100–1,500	1–2.7
For study		1,000	60	168	360	150	1,000	2.5
Paper/pulp (SIC-26)	10	500–2,500	15–100	168	250–360	30–150	1,500–2,000	10–30
For study		1,000	50	168	360	150	1,500	12
Textiles (SIC-22)	3	50–200	5–10	120–144	240–360	25–150	100–600	7–20
For study		150	6	130	360	150	200	15
Food (SIC-20)	4	5–180	0.4–12	120–168	280–345	50–125	100–1,000	1.5–25
For study		100	4	120	280	50	300	20

*Electric to total thermal ratio (each quantity in Btu/hr).

FIGURE 4.2: INDUSTRIAL ELECTRIC DEMAND COMPARED TO ELECTRIC-TO-THERMAL DEMAND

(Plant distribution and regression line)

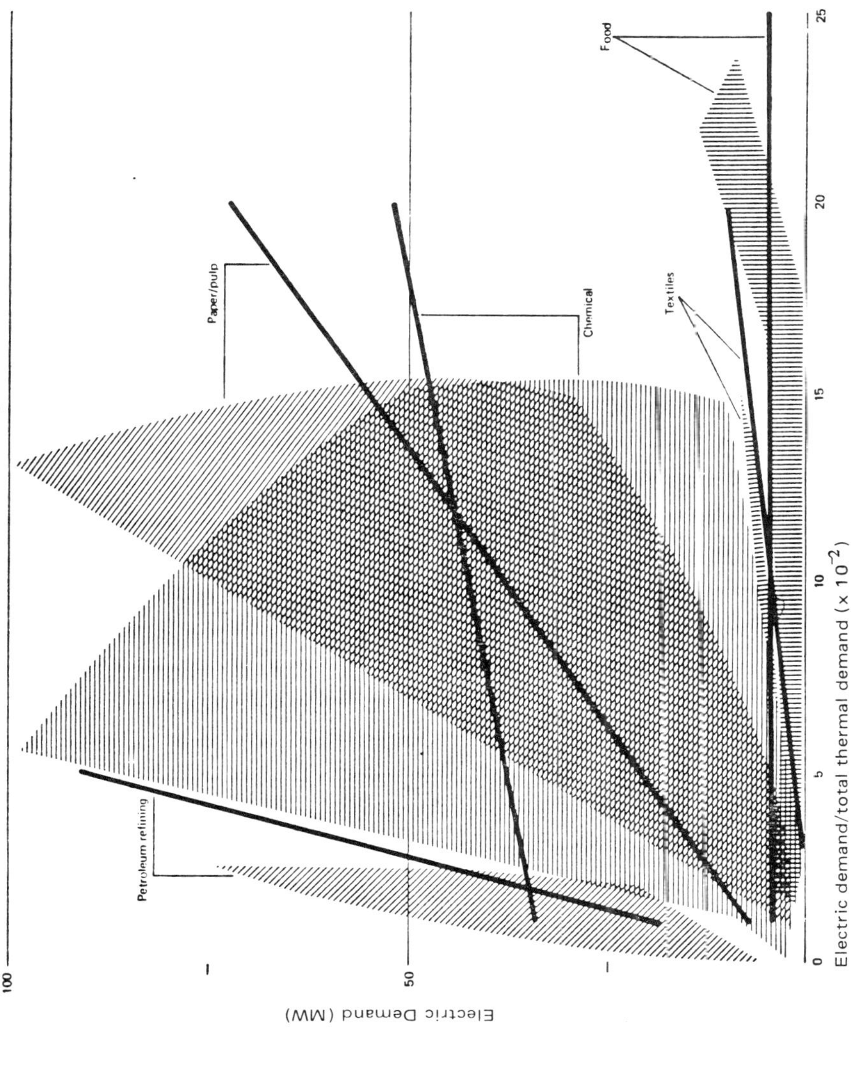

TABLE 4.2: WASTE HEAT PROFILES FOR 1977 (10^{12} Btu/yr)

Industry	Temperature range (°F) <150	150-300	300-600	>600	Total
Chemical	540	1,188	611	108	**2,447**
(percent)	22	49	25	4	**100**
Petroleum	147	428	247	280	**1,102**
(percent)	13	39	23	25	**100**
	<200	**200-500**	**>500**		
Paper/pulp	1,400	760	120		**2,280**
(percent)	61	33	6		**100**
Textile		180			**180**
(percent)		100			**100**
Food	800	160	120		**1,080**
(percent)	74	15	11		**100**

There are plants in almost every state, with the heaviest concentration of large plants in the West South Central Region which accounts for 16% of total sales. Other regions with large chemical complexes, but smaller plants, are the Middle Atlantic, East North Central, and South Atlantic which together account for 60% of sales.

TABLE 4.3: CHEMICAL INDUSTRY—FUEL AND PURCHASED POWER CONSUMPTION/1970

SIC Category	Chemicals in Group	Feedstock, Fuel, Purchased Power* (10^{12} Btu)	Percent of Total
281	Industrial chemicals (organic/inorganic)	2,585.5	80.6
282	Plastics/synthetics	346.4	10.8
283	Drugs	54.4	1.7
284	Soaps/detergents	41.7	1.3
285	Paints, varnishes	25.7	0.8
286	Gum/wood chemicals	6.4	0.2
287	Agricultural chemicals	64.2	2.0
289	Miscellaneous chemicals (preparations, carbon black)	83.4	2.6
Total		**3,207.8**	**100.0**

* Purchased power based on 34 percent theoretical fuel efficiency.

The chemical industry used more energy than any other industry in 1976 (3.2 quads). The first two major classifications together used 92% of the total fuel and purchased power for the chemical industry in 1976. Through interviews with plant managers, it was found that one cannot generalize about the ways in which electricity, process steam, or process heat are used in production. Table 4.4 shows the types of energy used by the chemical industry in 1972 and 1976.

TABLE 4.4: ENERGY INPUTS TO THE CHEMICAL INDUSTRY

Fuel	1972 (10^{12} Btu)	Percent	1976 (10^{12} Btu)	Percent
Natural Gas	1,569.6	47.2	1,310.4	40.9
Distillate Fuel Oil	30.5	0.9	54.3	1.7
Residual Fuel Oil	173.9	5.2	191.6	6.0
Coal	317.7	9.6	280.5	8.7
Electricity	696.8	21.0	769.1	24.0
Steam	133.8	4.0	144.5	4.5
Propane	17.9	0.6	13.1	0.4
Other Liquified Petroleum Gases	6.7	0.2	11.3	0.3
Coke	7.7	0.2	6.5	0.2
Other Gases	333.0	10.0	352.4	11.0
Other Liquids	26.8	0.8	60.2	1.9
Other Solids	10.0	0.3	13.9	0.4
Total	**3,324.4**	**100.0**	**3,207.8**	**100.0**

In 1976 these inputs were converted into electricity, process steam and process heat in the following pattern:

Energy	Percent	10^{12} Btu
Electricity	18.3	509
Process steam	51.2	1,427
Process heat	30.5	849
Totals	100.0	2,785

Of the five industries studied, the chemical industry has the greatest diversity of cogeneration applications. Some chemical plants have jointly produced steam and electricity for 80 years. Steam turbine-generator sets burning coal, oil, or gas and gas turbine-generator sets burning gas or oil have been employed for process steam topping. Waste heat recovery also has been used.

Plant surveys showed that only 13% of the electricity and only 11% of process steam are provided through cogeneration. A large number of chemical plants have steam flow demands below the level (70,000 pounds per hour) currently considered the minimum required for cogeneration. Also, some chemical plants are farther than five miles from any other industrial plant or utility, limiting the chemical plant's ability to sell steam.

These three factors constrain development of cogeneration in this industry. Load mismatch, power failures, and peak load usage further reduce the potential for topping of process steam by 15%.

Waste heat recovery equipment provided about 10% of the process steam used in the chemical industry in 1976. This heat is found primarily in cooling water and is transferred either to cooling towers or to natural bodies of water. The sources of waste heat are distillation condensers, compressor coolers, reactor cooling coils and product coolers. However, in a typical chemical plant, these sources of waste heat are small and widely dispersed.

The comparison of electric-to-thermal ratio with electric demand (see Figure 4.2) shows large dispersion within the industry. At one extreme, there are large-sized plants with slowly increasing thermal demand. Gas turbines and diesels are well-suited to satisfy these plants' electric-to-thermal requirements. On the other hand, for those plants that have small electric demands but a balanced thermal demand, the steam turbine would best meet their requirements. In general, most of the plants surveyed clustered in the range where gas turbines and diesel engines could be most attractive.

Petroleum Refining Industry

Petroleum refining (SIC 2911) is one of the most complex and technically sophisticated industries in the U.S. There are currently about 250 U.S. refineries, including fully integrated, highly complex plants, capable of producing a complete range of petroleum products and some petrochemicals; as well as very simple plants, capable of producing only a small number of products. Production ranges from a few hundred barrels per day (bpd) to over 400,000 bpd. Some refineries are modern, while others contain processing units constructed over 40 years ago. The types of crude oil refined also vary greatly.

The number of refineries in the U.S. has decreased in the last few decades, while the average size of a refinery has increased. Few new refineries have been built in the past five years; however, changes have been made in existing refineries to reflect changing technology and product demand. Although several new facilities are planned, many such projects have been either cancelled or greatly delayed, primarily because of unresolved environmental and energy issues. The rapid increase in costs has contributed to the decline in the construction of new refineries.

Although practically every refinery has its own capacity, processing configuration, and product mix, refineries in three regions (Northeast, South, or West) do tend to have some common characteristics. Refineries in the Northeast are the only ones that extract aromatics. They also refine more fuel through desulfurizing processes and catalytic reforming than refineries elsewhere. Southern (Gulf Coast) refineries produce more fuel through distillation than other refineries. Western refineries produce more fuel through hydrocracking and coking. Process heat is used primarily in crackers and reformers.

The distribution of input fuels is shown in Table 4.5. Since 1972, the petroleum industry has increased the efficiency of its use of energy by 12%. In 1976, these inputs were converted to electricity, process steam, and process heat in the following pattern:

Energy	Percent	10^{12} Btu
Electricity	2.2	74
Process steam	18.7	624
Process heat	79.1	2,641
Totals	100.0	3,339

TABLE 4.5: ENERGY INPUTS TO THE PETROLEUM INDUSTRY

Fuel	1972* (10^{12} Btu)	Percent	1976** (10^{12} Btu)	Percent
Crude Oil	2.9	0.1	2.8	0.1
Distillate Oil	17.4	0.6	27.0	0.9
Residual Oil	264.4	9.1	278.5	9.7
Liquified Petroleum Gas	29.1	1.0	42.4	1.5
Natural Gas	915.4	31.5	738.2	25.6
Refinery Gas	1,060.6	36.5	1,100.0	38.2
Petroleum Coke	427.2	14.7	433.1	15.0
Coal	5.8	0.2	4.7	0.2
Purchased Steam	2.9	0.1	38.2	1.3
Purchased Electricity	180.2	6.2	214.5	7.5
Total	**2,905.9**	**100.0**	**2,879.4**	**100.0**

* Estimated-based on a sample of 51 companies which operate 93% of current U.S. petroleum refinery capacity.

** Unadjusted for product mix and processing changes.

From plant surveys, it was found that only 7% of electricity and 3% of process steam are produced by cogeneration. Approximately 28% of total process steam is generated by waste heat recovery boilers. The types of cogeneration equipment presently used include steam and gas turbine-generator sets, both with waste-heat recovery boilers. However, due to load mismatches, the potential for topping of process steam is reduced by 10%.

Since the construction of new refineries has slowed, cogeneration will probably be introduced only when processes are modified to account for new fuels and products. For example, by 1985, natural gas will be phased out as a refinery fuel and tetraethyl lead as a gasoline additive.

Table 4.6 shows the pattern of energy use in the refining industry in 1974 and projected to 1985. Consumption of fuel and electric power is estimated to be about 18% higher in 1985. This increase will be due primarily to the upgrading of catalytic reformers to make unleaded gasoline.

TABLE 4.6: PATTERN OF REFINERY ENERGY USE

	East Coast 1974	1985	Gulf Coast 1974	1985	West Coast 1974	1985
Crude run, 10^3 BPD*	188	198	222	218	155	164
Purchased natural gas, BPD	2,500		5,400		6,390	
Refinery fuel, BPD						
Gas	4,810	6,990	5,450	6,880	7,190	7,500
Oil (blended)	7,210	9,640	5,350	10,700	2,140	11,690
Total fuel, BPD	**14,520**	**16,630**	**16,200**	**17,580**	**15,720**	**19,190**
Purchased steam, 10^6 lb/day	5.6	5.7	0.5	0.5	0.7	0
Electricity, MWh/day	713.7	838.7	744.2	875.1	766.8	897.3

*Barrels per day

The comparison of electric-to-thermal ratio with electric demand (see Figure 4.2) shows that as electric demand increases, thermal demand also increases, but not as fast.

The trend toward larger plants means that the demand for electricity will increase faster than demands for thermal energy. Furthermore, if recent trends in the energy mix are projected into the near future, an increase in the amount of coal used can be expected. On the basis of these trends, it is obvious that the petroleum refining industry could use cogeneration equipment with a high electric-to-thermal output ratio (i.e., gas turbines or diesel engines) that can burn coal derived products.

Paper and Pulp Industry

The paper, pulp, and allied products industry (SIC 26) encompasses the production and sale of pulp derived from wood and other fibrous raw materials; the manufacture of paper and paperboard products from pulp and waste; and the conversion of paper and pulp into end products such as boxes, writing paper, and sanitary tissue. It is the fourth largest purchaser of electricity and fuels (and the third largest user of water) among all U.S. industries. However, it is the only U.S. industry that uses significant amounts of its waste materials as fuel.

Papermaking and pulping are considered separately because the alternative pulping processes that can be used to make a given product differ considerably in energy requirements, but papermaking processes do not.

For example, about 3.5 times more energy is needed to pulp virgin fiber as is needed to pulp secondary fiber, but the energy used in making paper is about the same regardless of the kind of pulp used. Electricity and process heat are used for drying, heating and packaging in the paper and pulp mills. Also, the process steam and heat are used primarily in the chemical transformation of wood into pulp.

The amount of energy used in each of the major pulping processes is shown in Table 4.7. Both unit energy requirements and total annual use (which is a function of the amount of product made by each process) are shown. The total energy required to manufacture one ton of kraft pulp is significantly greater than that required for a ton of semichemical or mechanical pulp. However, the purchased energy requirement for kraft pulp is significantly less than for semichemical or mechanical pulp. Moreover, because kraft manufacture is by far the largest segment of the industry in terms of total production, it is also largest with respect to total annual purchased energy. Within the kraft pulping process, the integrated and market pulp segments of bleached kraft are the largest; other segments, notably unbleached kraft and dissolving kraft, are comparatively minor.

Table 4.8 shows the energy used in converting the intermediate product, pulp, into paper and paperboard products. Table 4.9 lists the total energy use by type. Between 1972 and 1976, the energy obtained from

TABLE 4.7: ENERGY USE FOR SLUSH PULP PRODUCTION/1974

	1974 Production ('000 tons/yr)	Unit Consumption (10^6 Btu/ton): Total Unit Energy*	Unit Consumption: Recovered Energy**	Unit Consumption: Purchased Energy	Annual Total (10^{12} Btu): Total	Annual Total: Recovered	Annual Total: Purchased
I Wood pulp							
(A) Chemical pulp–Kraft bleached and semibleached							
Integrated	10,800	27.0	15.0	12.0	292	162	130
Market pulp and dried transfers***	4,700	27.0	15.0	12.0	127	71	56
Dissolving pulp	800	32.0	17.0	15.0	26	14	12
Subtotal, bleached and semibleached	16,300	27.0	15.0	12.0	445	247	198
Unbleached	16,980	18.0	14.0	4.0	306	238	68
Total kraft pulping	33,280	23.0	15.0	8.0	751	485	266
(B) Chemical pulp–other sulfite							
Integrated	1,610	21.0	10.0†	11.0	34	16	18
Market pulp and dried transfers***	700	21.0	10.0†	11.0	15	7	8
Dissolving pulp	970	28.0	17.0	11.0	27	16	11
Subtotal, sulfite	3,280	23.0	12.0	11.0	76	39	37
Semichemical	3,790	15.0	1.0	14.0	57	4	53
Total other chemical pulping	7,070	19.0	6.0	13.0	133	43	90
(C) Mechanical pulp							
Groundwood (bleached and unbleached)	4,490	15.0	1.0	14.0	67	4	63
Defibrated/exploded and other	3,380	4.0	0	4.0	14	0	14
Total mechanical pulp	7,870	10.3	0	8.6	81	4	77
II Waste paper							
(A) Deinking	2,200	4.0	0	4.0	9	0	9
(B) Nondeinking							
Old corrugated containers	5,700	4.0	0	4.0	23	0	23
News, mixed, pulp substitute	6,300	3.0	0	3.0	19	0	19
Total waste paper	14,200	3.6	0	3.6	51	0	51
Total all wood pulp and waste paper	62,420	16.3	8.5	7.8	1,016	532	484

*Total unit energy based on estimated existing industry average 10% higher than mill requirements.

**The purchased fuel equivalent of steam and power generated from spent pulping liquors, bark and hogged wood. Recovery efficiency for existing industry average estimated at 90% of new mill efficiency.

***No drying included. Dried transfers refer to intracompany pulp shipments of market pulp.

†Recovered energy for sulfite mills based on heat recovery for 75% of existing paper guide and 100% of existing dissolving capacity.

Note: Information obtained from *Statistics of Paper and Paperboard,* American Paper Institute.

TABLE 4.8: ENERGY USE FOR PAPER PRODUCTION

	1973 Production (000 tpy)	Unit Consumption, (10^6 Btu/ton)			Annual Total (10^{12} Btu)		
		Steam & Extracted Power	Purchased Electricity	Total Unit Energy	Steam & Extracted Power	Purchased Electricity	Total Energy
Newsprint & uncoated groundwood	4,800	12.0	0.7	12.7	58	3	61
Printing and writing	12,000	14.0	0	14.0	168	0	168
Packaging & industrial converting	5,940	15.0	0	15.0	89	0	89
Tissue and toweling	3,900	11.0	1.5	12.5	43	6	49
Linerboard	13,200	13.0	0.5	13.5	172	7	179
Corrugating medium	5,260	12.0	1.0	13.0	63	5	68
SBS board	3,660	14.0	0	14.0	51	0	51
Recycled boxboard	5,540	14.0	4.0	18.0	78	22	100
Construction paper and board	5,130	10.0	4.0	14.0	51	21	72
All other paperboard	1,060	12.0	4.0	16.0	13	4	17
Dissolving pulp	1,770	8.0	0	8.0	14	0	14
Market pulp and transfers	5,400	8.0	0	8.0	43	0	43
Total paper, board, and machine dried pulp	**67,660**	**12.3**	**1.9**	**14.2**	**843**	**68**	**911**

waste wood materials increased from 41.6% to 44.9% of the total energy used.

TABLE 4.9: ENERGY INPUTS TO THE PAPER AND PULP INDUSTRY

Fuel	1972 Percent	1976 Percent
Purchased Electricity	7.3	9.6
Purchased Steam	1.7	1.6
Coal	17.4	18.4
Residual Fuel Oil	36.6	41.8
Distillate Fuel Oil	1.6	1.5
LPG	0.2	0.1
Natural Gas	36.3	28.4
Other Purchased		0.2
Energy Sold	(1.1)	(1.6)
Total	**100.0**	**100.0**

In 1976, these inputs were converted to electricity, process steam, and process heat in the following pattern:

Energy	Percent	10^{12} Btu
Electricity	11.8	201
Process steam	82.3	1,401
Process heat	5.9	100
Totals	100.0	1,702

The paper and pulp industry uses cogeneration more than any of the four industries studied. Traditional back-pressure steam turbine-generator sets and gas turbine-generator sets are the major types of cogeneration equipment presently used by the industry. Back-pressure turbines are supplemented by extraction turbines for peaking. Forty percent of the electricity and twenty percent of the processed steam used were cogenerated. From on-site interviews, it was found that the potential for topping of process steam is decreased by 20% due to load mismatch, power failures, and peak loads.

Both the back-pressure steam turbines and the gas turbines use high pressure boilers. However, at present, these boilers cannot co-fire wood waste and coal. More research is needed on field-erected and packaged boilers that have that capability. The industry presently uses a large portion of the thermal energy in its waste streams and there is, therefore, only limited opportunity for additional bottoming of waste heat. This can be seen from the relatively low end-of-process temperatures and pressures (see Table 4.2). However, future technical developments could make it possible to capture waste heat from liquid effluents.

Process load requirements for process steam and electricity are becoming more balanced. Steam demands are decreasing, due to energy conservation measures; and electric demands are increasing, due to the use of new manufacturing equipment. In addition, comparison of the electric-to-thermal ratio with the electric demand (see Figure 4.2) shows that as plant size increases, the thermal demand changes very little. Thus, for larger plants, gas turbines become more attractive.

Textile Industry

The textile industry (SIC 22) is extremely diversified, producing a variety of fabrics and fabric blends, as well as a multitude of end products including knit apparel, yarns, and carpets. Textile plants range from highly integrated manufacturing complexes that process natural and synthetic fibers into finished products, to small, nonintegrated contract plants (commission finishers) that process goods owned by other producers.

Nearly 90% of the U.S. textile plants are in the East and over 40% are in the Southeast. The rest are scattered throughout the country. The industry's basic raw materials are wool, cotton, and synthetic fibers. Wool accounts for a small and declining proportion of the materials used by the industry. Cotton is still a major raw material, but its use shows little growth. The use of manmade fibers is growing rapidly.

In transforming a fiber, the industry uses both wet and dry processes. Weaving and knitting are essentially dry processes. Wet processes include sizing, desizing, washing, dyeing, scouring, mercerizing, and bleaching. Both wet and dry processes are found in most types of mills. All plants surveyed operate intermittently. Table 4.10 presents an energy use profile for all segments of the textile industry.

The textile industry is the ninth largest industrial user of energy. Four three-digit SIC categories use 66% of the energy used by the textile industry: cotton weaving mills, synthetic weaving mills, knitting mills, and textile finishing plants (see Table 4.11). Data on energy and fiber inputs to production were submitted to the American Textile Manufacturing Association by reporting firms for 1972 and 1976. Although production was slightly higher in 1976 than in 1972, less energy was used.

TABLE 4.10: ENERGY INPUTS TO THE TEXTILE INDUSTRY

Fuel	1972 Percent	1976 Percent
Bituminous Coal	12.46	12.30
Anthracite Coal	.27	.19
Propane	.93	1.14
Gasoline	.06	.08
Kerosene	.04	.07
Distillate	2.65	4.22
Residual Fuel Oil	15.23	25.26
Crude	.04	
Natural Gas	38.87	25.55
Electricity	28.77	30.68
Other Energy	.68	.51

In 1976, these energy inputs were converted to electricity, process steam, and process heat in the following pattern:

Energy	Percent	10^{12} Btu
Electricity	33.7	90
Process steam	52.1	139
Process heat	14.2	38
Totals	100.0	267

Only 6% of the electricity and only 18% of the process steam are cogenerated. Small plants (5 MW or less) presently do very little cogenerating. The main type of cogeneration equipment found in textile plants is the back-pressure steam turbine-generator set. There is a recent trend to coal-fired boilers. Little waste heat is available for reuse since most of it is under 250°F (see Table 4.2).

Under present conditions, intermittent operation implies the need to adjust steam and electric use to maintain high plant efficiency during periods of low steam and electric demands. From discussions with managers, it was found that load mismatch reduces the amount of steam suitable for topping applications by 50%.

The comparison of electric-to-thermal ratio with electric demand (see Figure 4.2) shows that as plant size increases, the thermal demand increases but not as much as the electric demand. Steam turbine-gen-

TABLE 4.11: TEXTILE INDUSTRY: FUEL AND ELECTRIC ENERGY USE/1971

SIC Code	Industry Group	Purchased Fuel (10^{12} Btu*)	Purchased Electricity (10^{12} Btu**)	Total (10^{12} Btu)	Percent
221	Weaving mills cotton	30.02	57.01	87.03	16.12***
222	Weaving mills synthetic	31.39	58.52	89.91	16.65***
223	Weaving & finishing mills, wool	10.58	5.35	15.93	2.95
224	Narrow fabric mills	3.75	3.54	7.29	1.35
225	Knitting mills	48.79	32.77	81.56	15.11***
226	Textile finishing (except wool)	80.86	16.69	97.55	18.07***
227	Floor covering mills	29.68	9.68	39.36	7.29
228	Yarn & thread mills	20.13	57.89	78.02	14.45
229	Miscellaneous textile goods	23.54	20.23	43.77	8.11
22	**Total textile mill products**	**278.42**	**261.45**	**539.87**	**100.0**

* Converted from Census table on the basis of 1 kWh = 3,412 Btu

** Converted from Census table on the basis of 1 kWh = 10,500 Btu (energy required at the generating plant).

***Total energy use in these categories represents more than 66 percent of textile industry energy consumption.

erators are well-matched to the textile industry's present steam and electric demands. However, for greater energy conversion efficiency and reduction in utility loads, larger plants should consider other types of cogeneration equipment. Gas turbines or diesel engines for topping of process heat could generate a singificant portion of the electric demand. The industry is also investigating ways to recover rejected heat from dyeing and finishing operations.

Food and Kindred Products

Only the manufacturing sector is considered here. The food industry (SIC 20) uses 16.5% of all U.S. energy, the sixth largest amount of energy of any industrial sector in 1976.

The manufacturing sector uses steam for cooking, and electricity and heat for evaporating and drying. Most food plants do not operate continuously. They are on-line, on the average, two shifts a day, five days a week; the work schedule is longer from time to time. The range of products is diverse and the amount of energy used to produce them varies. For example, the largest energy-users are beet sugar refining, wet corn milling, and meat packing which, together, use 25% of the energy used for manufacturing food.

Beet sugar, wet corn milling, and meat packing remained the most intensive energy users from 1971 to 1974, but their relative positions changed, due to differing degrees of energy conservation, and differing amounts of new plant equipment. Table 4.12 presents the entire set of food processing categories along with their total and relative energy use.

The types of energy used in the industry changed significantly between 1958 and 1974 (see Table 4.13). Purchased fuel declined steadily from 92.0% of total energy used to 86.8%. The shift from coal and fuel oil to natural gas is evident up to 1971 and reflects the desirability and cost advantage of using natural gas over other fuels. Beginning in 1971, there was a slight shift from natural gas to fuel oil that reflects the difficulties of obtaining natural gas in some areas of the country. During the entire period, the use of electric energy increased as a percentage of total energy. Total energy use increased at an average rate of 2.7% per year from 1958 through 1971. From 1971 to 1976, total energy use decreased at an annual average rate of 2.3% per year.

In 1976, these inputs were converted to electricity, process steam, and process heat in the following pattern:

TABLE 4.12: RANK ORDER OF SEGMENTS OF THE FOOD INDUSTRY BY ENERGY USE/1972

SIC	Industry	Energy use rank 1974	Energy use rank 1971	1972* 10^{12} Btu
20	Food and Kindred Products			
2063	Beet sugar	1	2	88.7
2046	Wet corn milling	2	3	63.5
2011	Meat packing plants	3	1	87.3
2082	Malt beverages	4	8	50.1
2033	Canned fruits & vegetables	5	6	52.6
2026	Fluid milk	6	5	50.5
2051	Bread, cake & related products	7	7	54.2
2075	Soybean oil mills	8	9	43.0
2062	Cane sugar refining	9	11	35.9
2037	Frozen fruits & vegetables	10	12	25.8
2048	Prepared feeds, n.e.c.	11	4	43.7
2079	Shortening and cooking oils	12	15	30.4
2099	Food preparations, n.e.c.	13	13	32.7
2077	Animal and marine fats & oils	14	10	32.9
2022	Cheese, natural & processed	15	19	21.8
2023	Condensed & evaporated milk	16	16	23.0
2032	Canned specialties	17	21	21.2
2086	Bottled and canned soft drinks	18	14	30.3
2016	Poultry dressing plants	19	17	14.5***
2085	Distilled liquor	20	18	21.4
2031	Sausage & prepared meats	21	20	21.2
2047	Pet food	22	**	14.2
2061	Raw cane sugar	23	23	14.2
2065	Confectionery products	24	22	18.2
2038	Frozen specialties	25	**	12.2
2031	Flour & other grain mill products	26	24	13.8
2034	Dehydrated food products	27	28	10.0
2052	Cookies & crackers	28	25	13.3
2095	Roasted coffee	29	26	14.2
2083	Malt	30	29	9.4
2043	Cereal breakfast foods	31	30	8.3
2074	Cottonseed oil mills	32	33	7.0
2035	Pickles, dressings & sauces	33	31	7.5
2024	Ice cream and frozen desserts	34	34	6.1
2021	Creamery butter	35	27	8.5
2066	Chocolate & cocoa products	36	37	5.1
2087	Flavoring extracts & sirups, n.e.c.	37	32	7.9
2017	Poultry and egg processing	38	**	4.4***
2091	Canned & cured seafood	39	36	6.2
2084	Wines, brandy	40	39	3.5
2092	Fresh or frozen package fish	41	40	3.7
2076	Vegetable oil mills, n.e.c.	42	38	4.1
2044	Rice milling	43	41	2.4
2097	Manufactured ice	44	35	3.2
2098	Macaroni and spaghetti	45	43	2.4
2045	Blended and prepared flour	46	42	2.6
2067	Chewing gum	47	44	3.1

* Estimated from U.S. Department of Commerce, 1972 Census of Manufacturers, November 1975.

** Because of a change of definition in the four-digit SIC code energy use for these component industries was not reported in 1971.

*** w/o transportation

Energy	Percent	10^{12} Btu
Electricity	20.3	142
Process steam	50.4	352
Process heat	29.3	205
Totals	100.0	699

The manufacturing sector of the food industry generates all of its steam in-house. In 1976, about 6% of all process steam and 6.3% of all electricity were cogenerated. The diversity of the food industry makes it difficult to generalize about the potential for additional cogeneration. However, intermittent operation makes load mismatch an important deterrent. From plant surveys it was found that load mismatch reduces the potential for steam topping by 70%. Energy storage would help solve this problem.

TABLE 4.13: FOOD INDUSTRY: HISTORICAL ENERGY PROFILE

Fuel Type	1958	1962	1967	1971	1974	1976
Distillate oil, %	NA	4.0	6.6	7.4	8.0	6.9
Residual oil, %	NA	12.4	8.2	7.3	7.8	6.7
Total fuel oil, %	22.4	16.4	14.8	14.7	15.8	13.6
Coal, %	32.5	18.9	24.0	13.6	9.5	7.2
Coke and breeze, %	0.2	0.2	0.2	0.2	0.2	0.2
Natural gas, %	36.2	43.1	49.8	57.9	57.1	48.4
Other fuels, %	1.7	3.3	1.9	1.9	4.2	15.0
Total purchased fuel, %	93.0	91.9	90.7	88.3	86.8	85.0
Net electricity, %	7.0	8.1	9.3	11.7	13.2	15.0
Total purchased fuel and net electricity, %	100.0	100.0	100.0	100.0	100.0	100.0
Total purchased fuel and net electricity, 10^{12} Btu	766.1	802.5	899.6	1,031.3	958.8	887.0

The food industry is highly sensitive to power failures, as the entire production could be lost due to spoilage. Consequently, very reliable and stable sources of power and steam are required. Plant owners say that cogeneration hardware has been very reliable, but new hardware must meet higher load factors and have lower downtime.

The comparison of the electric-to-thermal ratio with electric demand (see Figure 4.2) shows that thermal demand decreases as the electric-to-thermal ratio increases. This trend suggests that gas turbines and diesels with their greater electric-to-thermal ratios could be more attractive. Appropriately-sized steam turbines would best fit those plants where the electric and thermal demands are in better balance.

COGENERATION SYSTEMS

The simultaneous production of electric and thermal energy requires equipment that burns fuel to produce shaft power (for mechanical work, generation of electricity, etc.) and heat. Heat or thermal energy can be used as dry heat for an industrial process or converted to steam. The equipment that makes the bulk of energy used in an industrial energy system is called a prime mover. Steam and gas turbines and diesel engines have been used as prime movers in industry since the 1920s. In recent years, most of these prime movers have operated on natural gas or petroleum fuels. The conventional industrial operation, in which thermal and electrical energy are generated separately, is shown in concept in Figure 4.3.

FIGURE 4.3: CONVENTIONAL INDUSTRIAL PROCESS

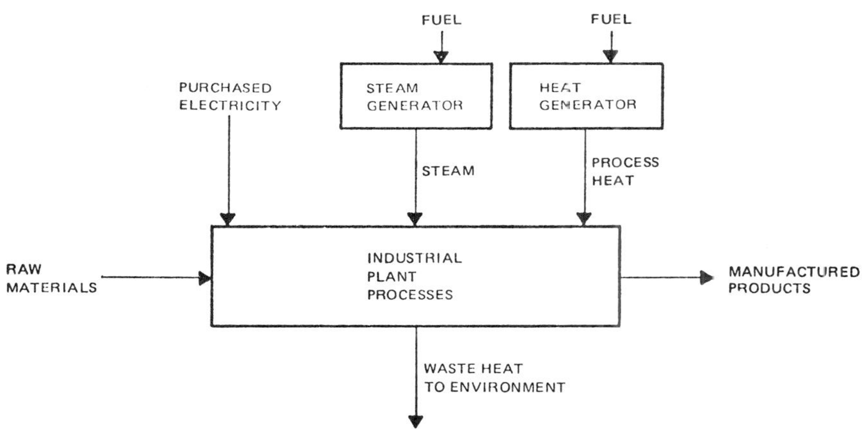

Steam turbine systems have been used to cogenerate steam and electricity since the turn of the century and are currently used in several industries, including the five of this study. Gas turbines have been used less for cogeneration in the past, but are now increasing in popularity because of their availability, small physical size, and flexibility in application. Until recently, the abundance of natural gas also favored the spread of gas turbines. Diesel engines have been used primarily for generating electricity for commercial buildings, hospitals, schools, and other small applications in this country, but are used extensively by electric utilities in Canada and Europe.

Electrical and thermal energy can be cogenerated by either or both of two techniques, topping of process steam or process heat or bottoming of processes, also called simply topping and bottoming cycles. In topping cycles, the by-product steam or heat from the prime mover (which is coupled to a generator) is then used for plant processes. Topping cycles are significantly more fuel efficient than conventional systems that generate electric and thermal energy separately.

Waste heat from plant processes can be recovered and converted, using bottoming cycles such as Rankine cycles to generate electricity. Waste heat recovery systems applied directly to gas turbine or reciprocating engine exhaust are also referred to as bottoming cycles. Bottoming cycles also save fuel when compared to conventional systems. Topping and bottoming cycles are shown schematically in Figure 4.4.

FIGURE 4.4: COGENERATION SYSTEMS

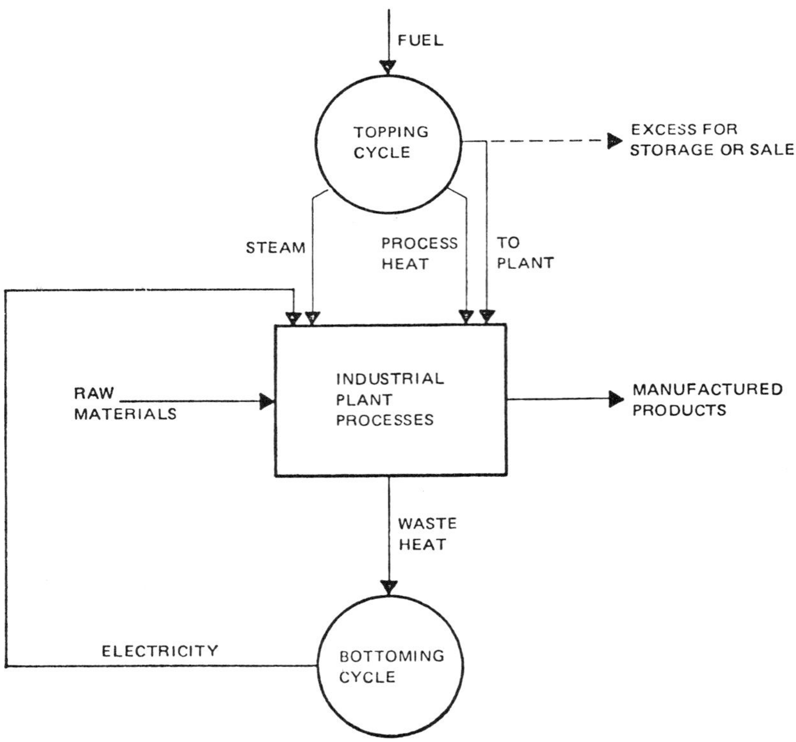

The most efficient prime movers for cogeneration, diesel engines and gas turbines, produce a higher ratio of electric-to-thermal energy (E/T)

than steam turbines. Diesels and gas turbines generate more electricity per unit of steam (see Figure 4.5) than the amount of electric power per unit of steam needed for most industrial plants. Therefore, to use the high efficiencies of diesels and gas turbines, the excess of electric power must be put to productive use, either by selling it to the utility grid, selling it to other industries, or storing it for later use. The export of surplus energy is not currently feasible due to institutional and regulatory barriers. Reducing these barriers would greatly increase the development of cogeneration.

FIGURE 4.5: TOPPING CYCLES: COMPARISON OF ELECTRICAL CAPACITY AND PROCESS STEAM CAPACITY

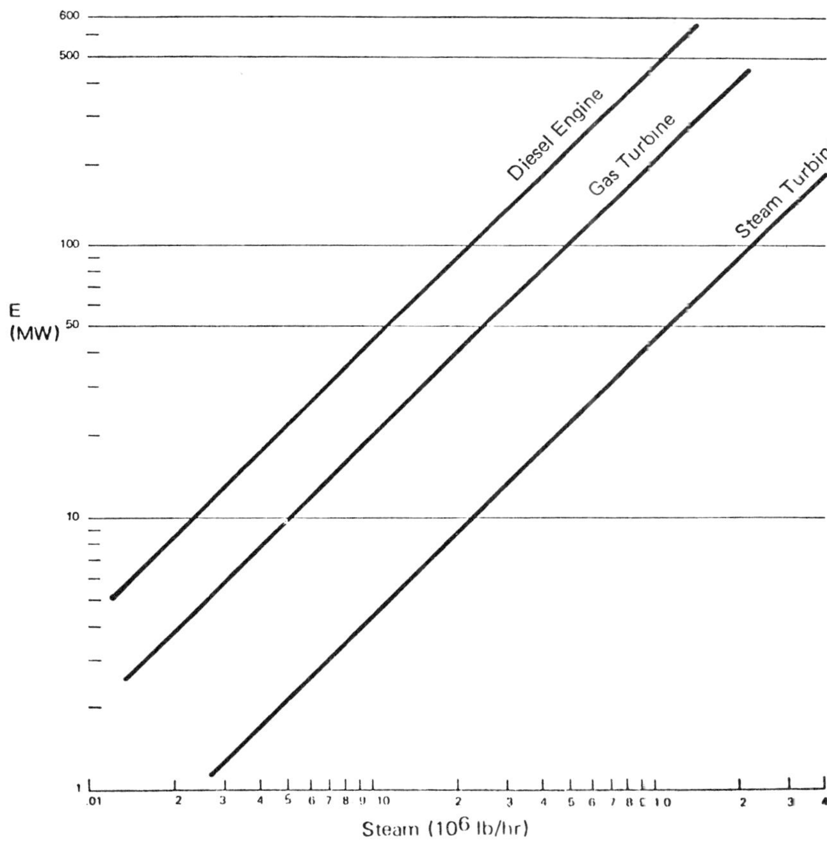

Current Technologies

Steam Turbine Topping Cycles: In recent years, the most common device for the topping of process steam has been the steam turbine. A fossil fuel-fired boiler is used to generate high pressure steam (e.g., 1000°F and 1,450 psig), which is then expanded through a steam turbine-generator set to produce electricity and low pressure process steam (see Figure 4.6).

FIGURE 4.6: STEAM TURBINE TOPPING CYCLE

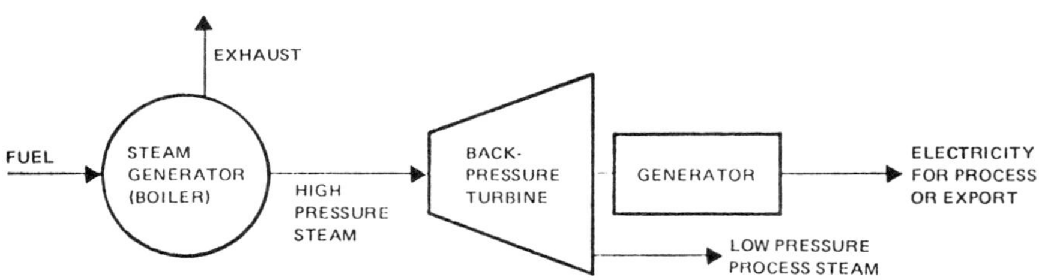

Back-pressure steam turbines (turbines that exhaust low pressure steam) are generally used for topping because they are most efficient. Multistaged steam turbines are also used since they permit the extraction of steam at any level of pressure below inlet conditions. Inlet steam throttle conditions can range from 250 psig saturated to 1,450 psig, 950°F steam. More electricity can be generated per unit of process steam under high throttle conditions, but the rate of fuel consumption per unit of process steam remains relatively constant over the range of throttle conditions. However, higher steam throttle conditions require slightly more expensive boilers, pumps, and water treatment equipment. The electricity-to-steam ratio under varying throttle conditions is shown below.

Throttle Condition	Electricity-to-Steam Ratio ($kWh/10^6$ Btu)
1,450 psig/950°F	47
850 psig/825°F	28
600 psig/750°F	24

Almost all industrial plants use packaged boilers for steam generation because they are cheaper and easier to install than field-assembled boilers. Most packaged boilers cannot be used for steam turbine topping because the pressure and temperature of the steam they generate are too low. Furthermore, they are normally less efficient than field-assembled boilers, and, unlike the latter, cannot usually be converted to burn coal economically. It is likely, therefore, that field-assembled, coal-fired boilers will be used more and more in industry. These boilers can be designed to burn coal, residual oil, and process wastes, and are suitable for steam turbine topping cycles.

Topping of process steam by a steam turbine requires less incremental fuel per unit of power generated (approximately 4,700 Btu/kWh) than topping with gas turbines or diesel engines, but it cannot produce as much electricity per unit of process steam. The ratio of electricity-to-steam is, however, generally suitable for industrial requirements.

Small steam turbines (less than 5 MW) are significantly less efficient and cost roughly three times as much per kW as large steam turbines. Small steam turbines could be used for topping in small industrial plants, however, combined gas/steam turbine cycles probably have a wider application. Research is now under way to improve the efficiency of small steam turbines and new designs are expected to be available commercially by the mid-1980s.

Increased use of steam turbines for the topping of process steam is not expected to have adverse effects on the environment except where coal is used as a fuel. Sulfur oxides are major pollutants from the burning of high sulfur coal. Fly ash is also a problem. Special emission control systems such as exhaust stack scrubbers will be required in most areas of the country where coal is burned.

Steam turbine technology has been available for several decades and reliability is high. Manufacturers usually recommend minor inspections annually and more detailed inspections every three years. Replacement parts are readily available and thus downtimes are usually short. Maintenance schedules need not be modified for varying throttle conditions or modes of extraction.

Gas Turbine Topping Cycles: The gas turbine is a more direct way to generate electricity than the steam turbine. Fuel, such as oil or natural gas, burned in the combustors of the turbine, heats air that has been previously compressed. The burned mixture expands through a turbine-generator set to produce electricity as shown in Figure 4.7.

Gas turbines operate at higher temperatures than steam turbines, and their exhaust can be recovered in a variety of ways. For instance, the exhaust steam can be passed through a heat exchanger (or boiler) transferring its heat to a process fluid such as water/steam, air, or oil. Since the exhaust gas is about 18% oxygen by volume, additional fuel can be fired in the exhaust stream to increase initial temperatures, as required by the process.

FIGURE 4.7: GAS TURBINE TOPPING CYCLE

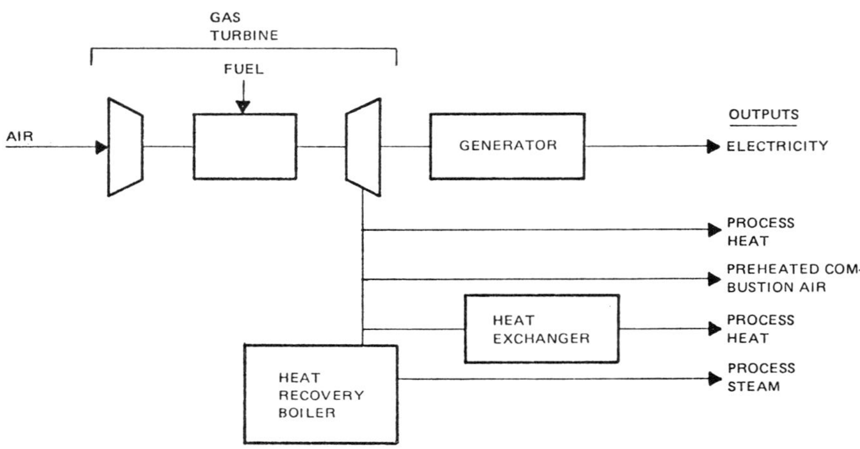

The exhaust stream, which is relatively clean and dry, can also be used directly for heating or drying, which is important in the food industry. If lower temperatures or increased volume are desired, the exhaust stream can be mixed with ambient air. If the back pressure of the process is low, a simple inductor is all that is required to provide the dilution air. At higher back pressures, a fan is required. As in the previous case, additional fuel can be burned in the exhaust prior to the process to match various temperature requirements.

The exhaust stream can be used for space heating or as highly preheated combustion air in boilers, oil heaters, hot gas generators, etc. This method is very efficient, since all the exhaust energy above the ambient temperature would contribute to fuel savings in the combustion device. This method has been used in the petroleum industry.

The last method is to pass the exhaust stream (e.g., at conditions of approximately 900 psig, 830°F), through a heat recovery boiler to generate process steam.

Current gas turbine power plants operate at thermal efficiencies well below 30% and topping cycles have heat rates of 5,500-6,000 Btu/kWh. Optimal performance is achieved at peak rather than base loads. Significant improvement in gas turbine performance as well as fuel flexibility are expected in the near future.

Best performance and least emissions are achieved if natural gas is used as fuel. Clean light petroleum distillates (such as naphtha) also provide acceptable performance and some gas turbines are currently available with fuel flexibility. These turbines can switch without interruption from natural gas to distillates.

New fuel treatment systems will enable gas turbines to use heavier oil distillates as well. These systems remove such trace metals as sodium, potassium, and vanadium (which cause corrosion or deposits) from the crude or residual oil. The contaminated oil is washed and the clean fuel separated by centrifuging or electrostatic separation.

Gas turbines can also be modified to run on low-Btu gas (approximately 100-150 Btu/scf. In Japan, a turbine has operated on blast furnace gas for 35,000 hours and another on coke oven gas for 27,000 hours. Gasification systems that can convert any commercial grade of crude or residual fuel to low-Btu gas are under development. Manufacturers recognize that future marketability will be limited unless gas turbines are adapted to burn fuels derived from coal or oil shale as they become available.

High temperature, high performance gas turbines are currently limited by the materials from which they are made. Normally, excess air must be supplied in the combustors to keep the temperatures of the blades down. One method that permits firing at high temperatures is the passage of cooling air through holes in the blades. Future turbines may use film cooling and transpiration cooling (in which the cooling air passes through pores in the airfoil and into the boundary layer to insulate the blade surface from hot gas).

Improved blade and nozzle materials, such as ceramics, can also allow higher firing temperatures. A major research effort is under way to use ceramics, principally silicon, nitride, and carbide, in gas turbines, diesels, and other equipment. Ceramics tolerate higher operating temperatures (in the range of 2500°F) resulting in significant increases in heat engine efficiencies. Gas turbine thermal efficiency could increase by 10 to 15%.

It must be pointed out that the current trends in gas turbine research may not be the most desirable. Current efforts are directed mainly at increasing performance. However, if increased operating efficiency results in higher capital cost, market development may suffer. The industrial managers interviewed felt that gas turbines needed greater reliability and fuel flexibility more than they needed improved performance.

Use of gas turbines for cogeneration is expected to increase. Petroleum refineries and chemical plants have recently installed them. However, significant market development will be unlikely unless their fuel flexibility and performance are improved. Unless they can operate efficiently on abundant lower grade fuels, gas turbines will not sell well until coal-derived fuels are available. Thereafter, market development will depend mostly on economic factors and the availability of better energy storage systems. The primary demand is expected to be in the 5 to 100 MW range.

The effects on the environment of increased industrial gas turbine topping cycles will depend on the fuels they burn. EPA requirements will limit NO_x emissions and the sulfur content of fuels. Better combustors are needed to meet new emissions levels with all fuels except lean distillates and natural gas.

Gas turbines require more frequent maintenance than steam turbines. The combustion system must receive frequent minor inspections; and bearings, rotors, etc., must be inspected at intervals ranging from 5,000 to 10,000 hours, depending upon the type of fuel and the frequency of starting. Major overhauls are required at 20,000 to 75,000 hours. The complete system or its major modules can be replaced without major disassembly on-site.

Combined Gas/Steam Turbine Topping Cycles: Combined gas/steam turbines provide greater flexibility for industrial topping of process steam than either simple gas or steam turbines. Their performance, however, can only duplicate that of separate gas turbine topping or steam turbine topping. Combined gas/steam turbine topping cycles can include special heat recovery boilers that allow the gas turbine exhaust to be used as combustion air, as shown in Figure 4.8.

Additional fuel is burned in the heat recovery boiler to produce high pressure steam that is then expanded through a back-pressure steam turbine to produce low pressure process steam and electricity. The terms unfired, supplementary fired, and fully fired refer to the addition of varying amounts of fuel to the heat recovery boiler for additional energy. Because of its similarities with gas turbine and steam turbine topping, combined cycle topping was not investigated further in this study.

FIGURE 4.8: COMBINED GAS/STEAM TURBINE TOPPING CYCLE

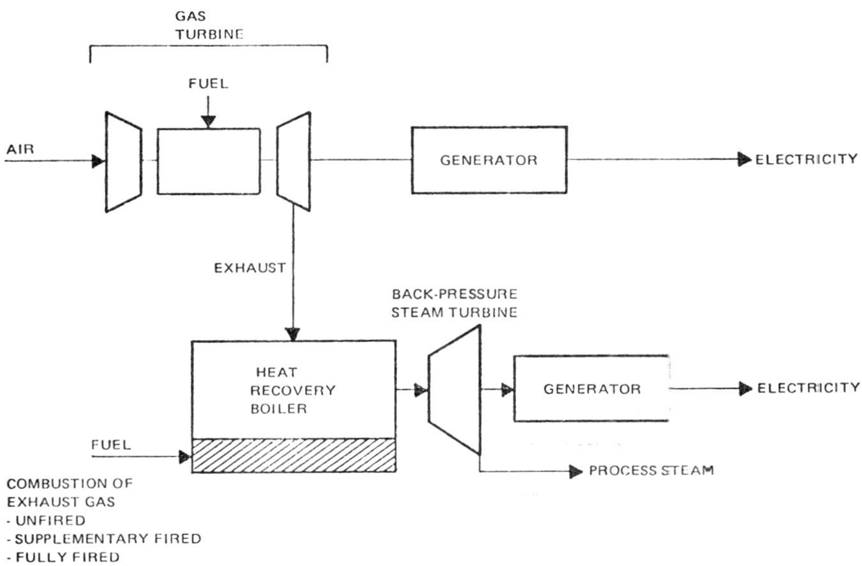

Diesel Enginge Topping Cycles: Topping of process steam or process heat with a diesel engine provides the highest electricity-to steam ratio and is available in a range of sizes from 100 kW to 25 MW. Basic thermal efficiencies are approximately 35%, compared to less than 30% for gas and steam turbines. Heat rates are in the 6,000-7,000 Btu/kWh range, less favorable than for the other prime movers. However, electricity generation per unit of steam is twice that of gas turbines and more than 10 times that of steam turbines.

The concept of diesel topping is shown in Figure 4.9. The diesel engine drives an electric generator. Process steam is made in heat recovery boilers that use the engine exhaust and jacket cooling as heat sources. Diesel engines can be coupled to bi-phase bottoming cycles or Rankine bottoming cycles or to generate additional electricity.

FIGURE 4.9: DIESEL ENGINE TOPPING CYCLE

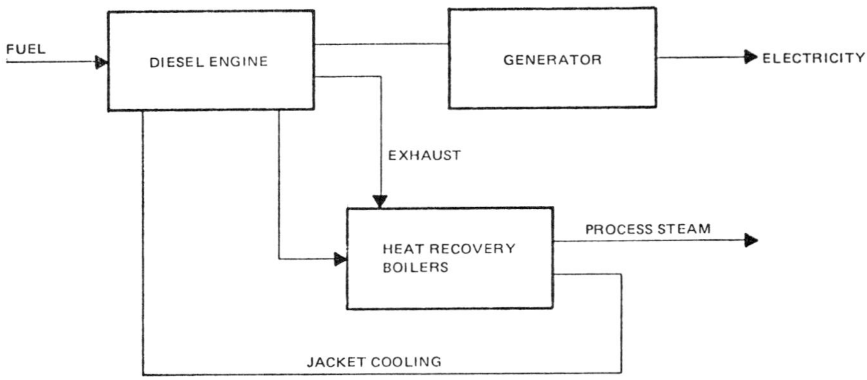

Diesels currently depend on petroleum-based fuels. They can be designed to use both high distillates and low-grade crude and residual oils. Filtering and centrifuging eliminate contaminants that may cause excessive wear in the injection system and in the engine's moving parts. Research is under way on a dual-fuel diesel that can use both liquid and gaseous fuels such as propane, butane, and methane. Research has also started on the use of coal-derived fuels; these will be essential to the future market development of diesels.

Ceramic pistons, cylinder liners, and heads would offer other potential benefits, such as improved efficiency, decreased fuel consumption and extended maintenance periods for diesels. Experiments on the use of ceramics in diesels have been conducted in Sweden and are beginning in this country. Ceramics can provide long life at higher pressures and temperatures.

The spread of diesel topping systems will be restrained by two major factors: first, the electricity-to-thermal ratio of diesel engines is significantly higher than most industrial plants need. Diesels will not be economically attractive unless excess electric power can be sold or stored. Second, diesels will also have to convert to more abundant coal-derived fuels.

Diesel-generator sets greater than 25 MW have not been commercially available in the U.S. Large "cathedral" diesels have been used in Europe for marine propulsion and generation of electricity. The demand for large diesels may develop in this country with increased industrial incentives for cogeneration. The size of the equipment lessens the advantages of diesel units where space is a premium. Advanced, more efficient units will be ligher and smaller, and will alleviate this problem somewhat.

The environmental impact of diesel topping cycles is minimal if light distillate fuels are burned. Ceramic components will improve combustion and lower exhaust emissions. If coal is used as a fuel, sulfur oxides must be removed from the stack gas to meet emission standards.

The reliability and service life of diesel engines is high. They typically require minor maintenance at 7,000 to 10,000 hour intervals and major overhauls every 20,000 to 30,000 hours. Ceramic components, which would reduce wear on rings and cylinder liners, should extend service life.

Steam and Organic Rankine Bottoming Cycles: Steam and organic Rankine-cycle bottoming systems convert waste heat to electric and useful thermal energy. In the Rankine system, a liquid working fluid is heated and vaporized in a vapor generator at constant pressure. The hot vapor is expanded to a lower pressure in an expander, producing electric or shaft power. The turbine exhaust is then cooled and condensed at constant pressure. The condensed fluid is pressurized by a pump, preheated, and returned to the vapor generator (see Figure 4.10).

FIGURE 4.10: RANKINE BOTTOMING CYCLE

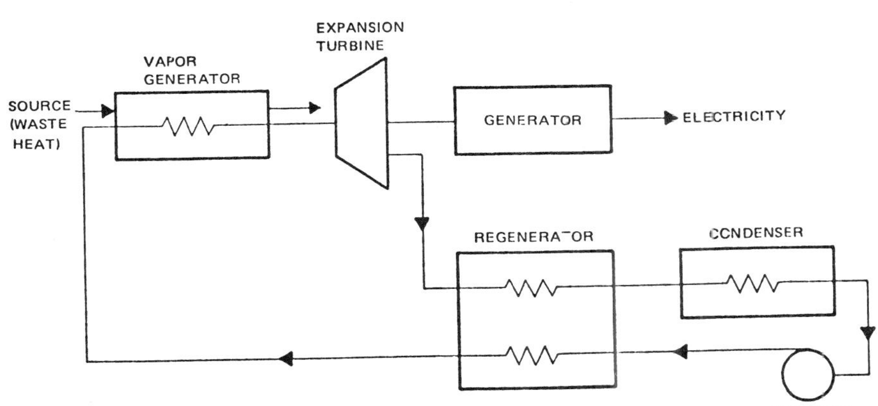

Bottoming cycles can be used to recover waste heat from many industrial processes and from diesel engine exhaust and jacket cooling water. Practically all of the Rankine cycle systems currently in commercial use employ steam as the working fluid. They are used for a wide variety of mechanical drive and electrical power generating purposes. These systems have thermal efficiencies from 14 to 36%, operate over a temperature range of 400° to 1000°F, and are greater than 670 hp (approximately 500 kW) in size.

New Rankine cycle systems using organic working fluids are commercially available in a limited number of sizes, ranging from 3,000 to 5,000 hp (approximately 2,000 to 3,800 kW). These operate at much lower temperatures (200° to 340°F) and so, despite their lower efficiences (less than 13%), can recover waste heat more effectively.

The choice of a working fluid for Rankine cycle heat engines depends on such factors as the temperature profile of the waste heat and the output power level. Research is being conducted on the use of organic fluids at operating temperatures up to 1000°F and efficiencies up to 20%. System characteristics can be improved through further development of fluids, bearing, and seals.

New working fluids or mixtures of existing fluids might reduce the need for superheating and desuperheating, thereby increasing efficiency, reducing heat transfer area, and providing chemical stability at high operating temperatures.

The effects of bottoming systems on the environment include the reduction of thermal pollution of the atmosphere. However, organic Rankine systems working fluids are generally toxic and therefore pose an environmental risk.

The maintenance requirements and reliability of bottoming cycles are generally unknown because there is little commercial experience with these systems to date. However, the technology of the component parts (expansion turbines, fluid pumps, heat exchangers, etc.) is proven and since the systems operate at lower temperatures than conventional systems, reliability is expected to be high.

Near State-of-the-Art Technologies

Fluidized-bed combustion systems and bi-phase turbine cycles are presently under full scale demonstration and will be commercially available by the early 1980s. Fluidized-bed combustion has several advantages that will enhance industrial topping cycles. Bi-phase turbines show promise for bottoming applications as a lower cost alternative to organic and steam Rankine systems.

Fluidized-Bed Combustion Systems: Fluidized-bed combustion (FBC) is an alternative to the conventional water-tube industrial boiler. Although FBC has not yet gained commercial acceptance, several advantages of the technique indicate widespread use in energy-intensive industries by 1985. There are three basic FBC schemes applicable to cogeneration: atmospheric pressure FBC, pressurized combined cycle FBC, and pressurized combined cycle FBC with adiabatic combustion.

Both atmospheric and pressurized cycle FBC techniques can be readily integrated into steam turbine and gas turbine cogeneration systems. The advantages of the FBC over conventiona boilers are:

1) Ability to use compliance and noncompliance (e.g., high sulfur, etc.) coal as well as other low grade fuels, e.g., wood scraps, residual oil, peat, etc.;
2) Effective direct capture of sulfur dioxide from the combustion of coal;
3) Significantly lower NO_x emissions than conventional combustion;
4) Relatively unobjectionable solid wastes; and
5) Ability to be shipped in modules for simple field assembly, even for larger units.

Atmospheric fluidized-bed combustion occurs at normal atmospheric pressure in the temperature range of 1400° to 1500°F with excess air values of 15-25%. The bed consists of a mixture of crushed limestone, dolomite, or inert material, and large ash particles, which is "fluidized" by a stream of air or combustion gases rising from the supporting grid beneath the bed. Steam is produced in tube bundles or water walls located in the fluidized region. Conventional steam turbines convert this steam to electrical energy (see Figure 4.11).

Pressurized combined cycle FBC uses a steam turbine and a gas turbine. Combustion occurs in a fluidized-bed unit at temperatures of 1600° to 1800°F under pressure of 4 to 10 atmospheres. This technology reduces the size, and, thus, the cost of the combustion unit. The hot, high pressure fuel gas, existing from the combustion unit is expanded through a gas turbine to generate electric power. However, conventional steam turbines, driven by steam produced in the bed tubes, produce approximately 80% of the total power generated (see Figure 4.12).

FIGURE 4.11: ATMOSPHERIC FLUIDIZED BED COMBUSTION

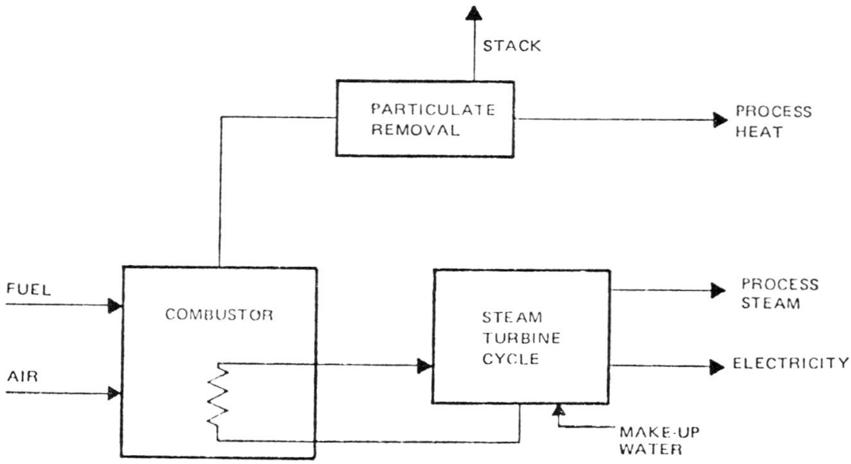

FIGURE 4.12: PRESSURIZED, COMBINED CYCLE FLUIDIZED-BED COMBUSTION

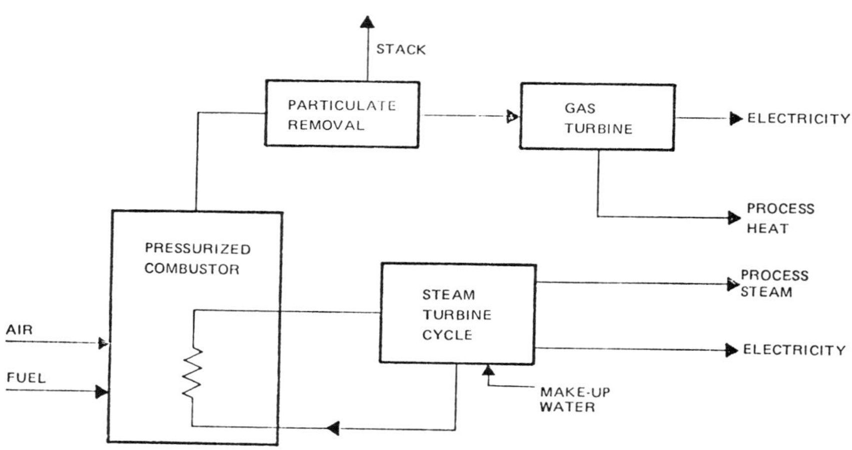

A pressurized combined cycle FBC unit with adiabatic combustion has no water tubes in the combustor. A large volume of combustor air (approximately 300% excess air) at a pressure of 4 atmospheres is expanded through a gas turbine for generation of electric power. A low pressure steam turbine bottoming cycle also generates power. Of the total output, approximately 80% is due to the gas turbine and 20% to the steam turbine (see Figure 4.13).

FIGURE 4.13: PRESSURIZED, COMBINED CYCLE ADIABATIC FLUIDIZED-BED COMBUSTION

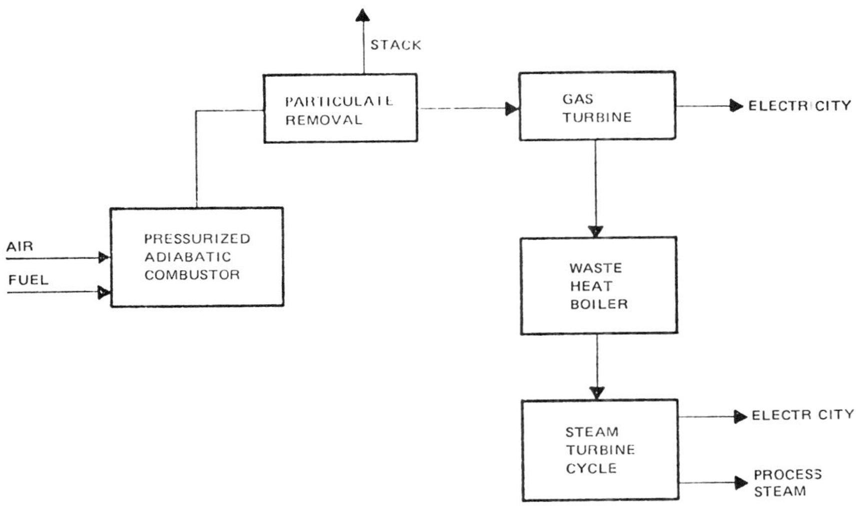

Several large commercial FBC boilers are expected to be in operation by the late 1970s. Pressurized FBC technology is somewhat further behind, with commercial operation expected in the mid-to-late 1980s.

FBCs are expected to spread quickly because they can burn coal, including high sulfur coal, within acceptable emission standards. FBC has excellent potential for cogeneration since gas turbines and steam turbines can be used to make process steam, electricity, and process heat simultaneously.

Bi-Phase Turbine Bottoming Cycle: The bi-phase power system is similar to a Rankine bottoming cycle, but has several potential advantages. The bi-phase system uses a mixture of liquid and gas as a working fluid to drive a specially equipped turbine. The system features a two-phase nozzle and a rotary separator turbine that extracts work from a vapor containing a finely dispersed liquid. The high mass flow rate and low spouting velocity realize maximum efficiency at low turbine speed. A single stage bi-phase turbine achieves as much net output power at approximately 8,000 rpm as a four-stage organic Rankine turbine at 15,000 rpm or a four-stage steam turbine at 26,000 rpm.

Slower turbines can use cheaper bearings, seals, etc. The bi-phase system has a gas-to-liquid heat exchanger instead of a boiler and thus more exhaust energy is available for conversion. The available energy is greater because heat transfer is to the liquid phase. Variations of the bi-phase power cycle may allow the use of a contact heat exchanger which is cheaper than a tube-fin heat exchanger.

The bi-phase turbine may be up to 50% less expensive than organic or steam Rankine bottoming systems in the operating temperature range of 450° to 1000°F. The advantages of the bi-phase turbine could result in early commercialization.

Components of the bi-phase turbine bottoming cycles are well within the state-of-the-art. Reliability is expected to be high. The cycle will reduce both air and thermal pollution.

Future Technologies

The techniques of cogeneration discussed in previous sections of this report have the potential for saving large amounts of energy. Improvements in their reliability and flexibility will broaden their application. Greater fuel flexibility will make them even more attractive to industry. There are, however, other technologies, still in the earliest stages of development, that show promise of eventually finding applications in industrial cogeneration. Among the most promising of these new technologies are improved energy storage, improved Stirling cycle engines, the Lysholm expander, and advanced heat pumps.

Energy Storage Systems: Gas turbines and diesel engines are theoretically the most efficient devices for topping. However, they have a much higher electric-to-thermal ratio than is needed by most industrial processes and they are not truly efficient unless all the energy they generate can be used. Surplus electricity could be exported from the plant, but, in many cases, institutional, regulatory, and technical barriers presently preclude this option.

Such barriers may eventually be lowered, but improved energy storage might be another way to solve the problem of surplus electricity. Improved storage systems would also increase overall system reliability. There are three main objectives for energy storage: to improve the salability of electricity by decreasing fluctuations; to level loads in-plant (short term storage); and to use in-plant at some future time (long term storage). At present, research on storage is principally being performed by public utilities and federal agencies.

Electricity can only be stored if it is converted to some other form of energy—chemical, mechanical, thermal, etc. Current technologies for storing energy are summarized in Table 4.14.

Compressed air and pumped hydro are the cheapest forms of storage for large facilities (200-2,000 MW). Lead acid batteries can be used by smaller facilities, and as they improve (between now and 1985), they will be suitable in the 20 MW to 50 MW range. New designs and materials may make flywheels a good form of storage in the 20 MW to 50 MW range as well. Between 1985 and 2000, advanced batteries should raise efficiencies and lower costs considerably. Thermal storage has a longer expected life, but is more expensive. Production and storage of hydrogen will be the least efficient and most expensive technique.

Stirling Engine: The Stirling engine is an externally fired, multifuel, high efficiency closed cycle heat engine. Although the Stirling engine has had thermodynamic efficiencies of up to 39% since the late 1950s, only relatively small lightweight units for mobile power applications have been developed. Because of their potential efficiency and durability, Stirling engines are now being developed for stationary power applications.

Large Stirling engines are expected to have efficiencies in the range of 40 to 45%. They are expected to be clean and quiet and to operate at low speeds. They can use a variety of fuels (or thermal sources), including petroleum distillates, shale or coal distillates, methane, low Btu gas, powdered coal, waste heat, and solar energy.

Stirling engines can switch from one fuel to another without adjustment, interruption of operation, or loss of power or efficiency. The moving parts of this external combustion system are not exposed to extensive erosion because of the low gas speeds in critical parts of the system. In addition, the Stirling engine can use almost any external heat source with temperatures greater than 1000°F to drive the engine. It could be used for cogeneration as an adjunct to fluidized-bed combustion and waste heat recovery systems. It is estimated that cogeneration systems using Stirling engines as the prime mover could save approximately 65% of the energy used in conventional generation systems.

Stirling engines have a relatively pollution-free exhaust. The NO_x content of emissions can be further reduced by recirculating part of the

TABLE 4.14: EXPECTED TECHNICAL AND COST CHARACTERISTICS OF SELECTED ENERGY STORAGE SYSTEMS

	Near Term					Long Term		
			Thermal					
Characteristics	Hydro Pumped Storage	Compressed Air	Steam	Oil	Batteries Lead Acid	Advanced	Flywheel	Hydrogen Storage
Commercial availability	present	present	before 1985	before 1985	before 1985	1985-2000	1985-2000	1985-2000
Economic plant size (MWh or MW)	200-1,000 (MW)	200-1,000 (MW)	50-200 (MW)	50-200 (MW)	20-50 (MWh)	20-50 (MWh)	10-50 (MWh)	20-50 (MW)
Power related costs* ($/kWh)	90-145	120-150	150-240	150-240	70-120	60-70	90-120	400-500
Storage related costs* ($/kWh)	2-10	3-10	35-70	10-12	50-60**	20-40	80-150***	6-10
Expected life (years)	50	20-25	25-30	25-30	5-10	10-20	20-25	10-25
Efficiency† (percent)	70-75	†††	65-75	65-75	60-75	70-80	70-85	40-50
Construction lead time (years)	8-12	3-12	5-12†††	5-12†††	2-3	2-3	2-3	2-3

*Constant 1975 dollars, does not include cost of money during construction.
**Could be considerably higher if buildings required to house batteries.
***These numbers are very preliminary.
†Electric energy out to electric energy in, in percent.
††Heat rate of 4,200 to 5,500 Btu/kWh and compressed air pumping requirements from 0.58 to 0.80 kWh (in)/kWh (out).
†††Long lead time includes construction of main power plant.

Note: Data from Schneider, Public Service Electric and Gas Company, presented at workshop on Lead Acid Batteries, Nov. 1975.

exhaust gas. The absence of explosive combustion, the smooth transition between maximum cylinder pressures, and the rhombic drive mechanism that establishes balanced crankcase conditions, contribute to the relatively quiet operation (typically 35 to 50% quieter than a comparable diesel engine).

It is not anticipated that industrial Stirling engines will become commercially available before 1985. Economics, fuel availability, and environmental factors will dictate the commercial prospect for Stirling engines. Stirling engines are expected to have operational lives of over 20 years. The need for long-term investment could discourage early commercial commitment; however, longevity accentuates the overall economic advantage. Multifuel Stirling engines might be considered insurance against interruptions of petroleum supply.

Lysholm Expander: The Lysholm expander is a high temperature (greater than 2500°F), helical expansion device made of ceramic materials. It will be developed for industrial cogeneration and topping cycles for central power plants.

The operating speeds and stresses encountered in the helical expander are an order of magnitude less than in conventional turbines. Thus, it can use durable, corrosion resistant ceramic components at high temperatures.

The expander may be driven by a direct, coal-fired Brayton topping cycle which would result in an overall thermal conversion efficiency exceeding 50%.

Primary application areas for the Lysholm expander are high temperature topping cycles for combined cycle electric power generation and alternatives to conventional turbines where contaminated combustion products must be tolerated. Combined with a standard steam electric power plant cycle, a 50 MW topping cycle unit could increase the energy conversion efficiency by 20%.

Ceramic helical expanders are especially tolerant of a variety of fuels. They can use solid, liquid, or gaseous fuels which can be derived from coal or other sources. Air pollution must be checked by various control technologies to remove fly ash, and SO_x. Exhaust gas scrubbers and precipitators will be needed for emission control as required.

The commercial potential of Lysholm expanders presently is inhibited by the materials limitation on operating temperatures, pressures, steam kinetics, etc. Assuming successful progressive development of high temperature ceramics and satisfactory solution of fabrication and design problems, the devices are expected to be commercially available by 1990.

Advanced Heat Pumps: Heat pumps can be used in industry to elevate the temperature of thermal energy in waste streams. The heat pumps can therefore be used in numerous industrial applications to reduce thermal demands by recovering and transforming low quality waste heat to usable levels.

Federally funded RD&D is presently aimed at improving component efficiencies (compressors and compact heat exchangers), lowering costs, and demonstrating reliability, specifically for use in the temperature range of interest for industrial applications.

Advanced heat pump projects being developed (including prototype design and performance evaluation tests for water-to-air and water-to-water systems), are expected to have a coefficient of performance (COP) of 3.5 to 5.0. (The COP is the ratio of heat output to power input to the compressor.) Present systems have demonstrated performance levels up to 3.0. This further increase in performance will make the heat pump a more attractive system. Ultimately, the widespread adoption of the heat pump will depend on how the marginal cost of recovering the available waste heat compares to the marginal cost of conventional supplemental heating techniques.

Advanced high temperature heat pumps are expected to become commercially available in the 1980s and achieve a reasonable degree of penetration in the industrial, commercial and residential markets by 1985.

APPLICATIONS

Topping Cycles

The choice of specific cogeneration technology and hardware for topping can only be made on a plant-by-plant basis. This is due to plant requirements for thermal and electrical energy, availability of fuel and space, and other site-specific characteristics. Steam and gas turbines and diesel engines differ in thermal and electrical output capabilities.

The most energy is saved if the equipment is designed to meet in-plant thermal demands and excess electrical energy is exported. Thus, energy savings are directly related to the quantity of electrical energy exported.

In Figures 4.14 through 4.23, the demands for process steam and process heat are compared with the demand for electric energy for specific chemical, petroleum refining, paper and pulp, textile, and food processing plants. Each point represents an average daily demand for process steam or process heat and electricity. These exhibits also show estimated performance curves for steam turbine, gas turbine and diesel engine topping techniques. The performance curves represent an average of manufacturers' stated output performance characteristics. Assumptions were made regarding steam pressures and temperatures and process heat temperatures for each industry application (see Table 4.1).

These exhibits show which technology (steam turbines, gas turbines, or diesels), if any, has electric and process steam or process heat outputs that best fit the demands of the various plants shown. In most cases, diesels and gas turbines produce more electricity, per unit of process steam or process heat, than the plant can use. The exhibits also show the size of the surplus for any particular plant. For example, assume power generation equipment was designed to meet a plant's steam demand of 0.6×10^6 lb/hr (see Figure 4.14). A steam turbine would generate only 25 MW of electric power, which is 15 MW less than the plant's electrical demand (40 MW). A gas turbine, however, would generate approximately 125 MW of electricity, yielding a 65 MW surplus. A diesel engine would generate approximately 290 MW, yielding a surplus of 250 MW.

Steam turbine topping best fits most plant needs in all five industries. Gas turbines best fit the process heat demands of most plants in all the industries except paper and pulp and textiles. Diesels generate the most electricity and also the greatest surplus of electricity that must be stored or sold to the utility grid.

Bottoming Cycles

Various bottoming cycles for the five industrial groups were evaluated. Only the chemical and the petroleum refining industries had sufficient recoverable waste heat for bottoming applications. Waste heat profiles of these two industries were applied to the operating performance curves of each bottoming cycle (see Table 4.2 and Figure 4.24).

FIGURE 4.14: TOPPING OF PROCESS STEAM
(Comparison of Electrical Capacity and Steam Demand in the Chemical Industry)

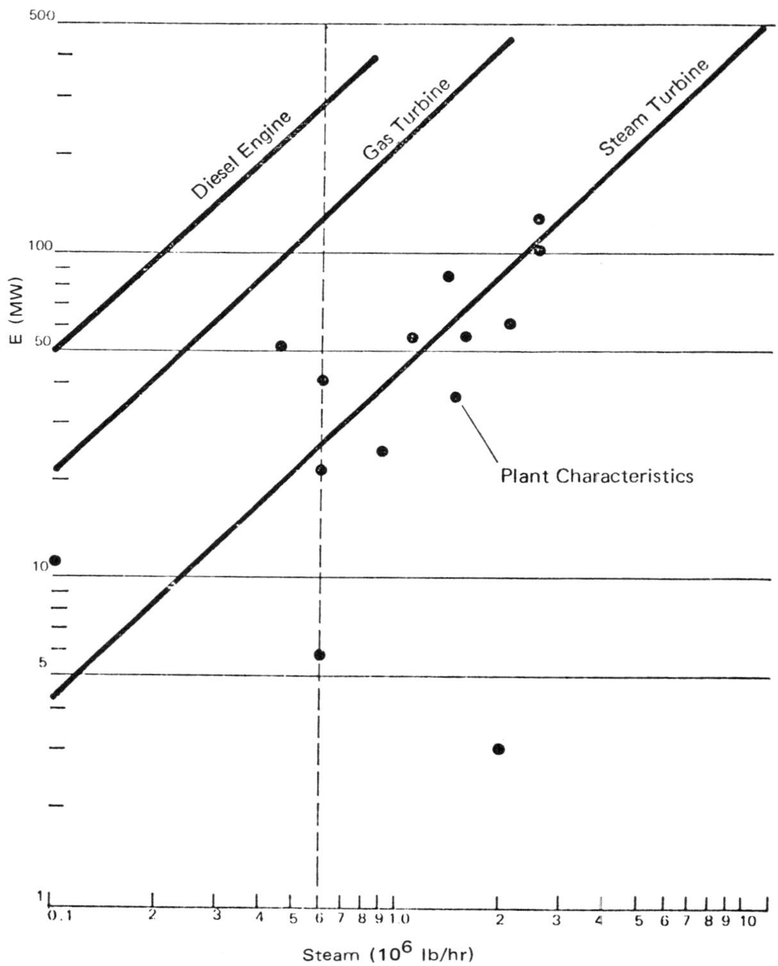

FIGURE 4.15: TOPPING OF PROCESS STEAM

(Comparison of Electrical Capacity and Steam Demand in the Petroleum Refining Industry)

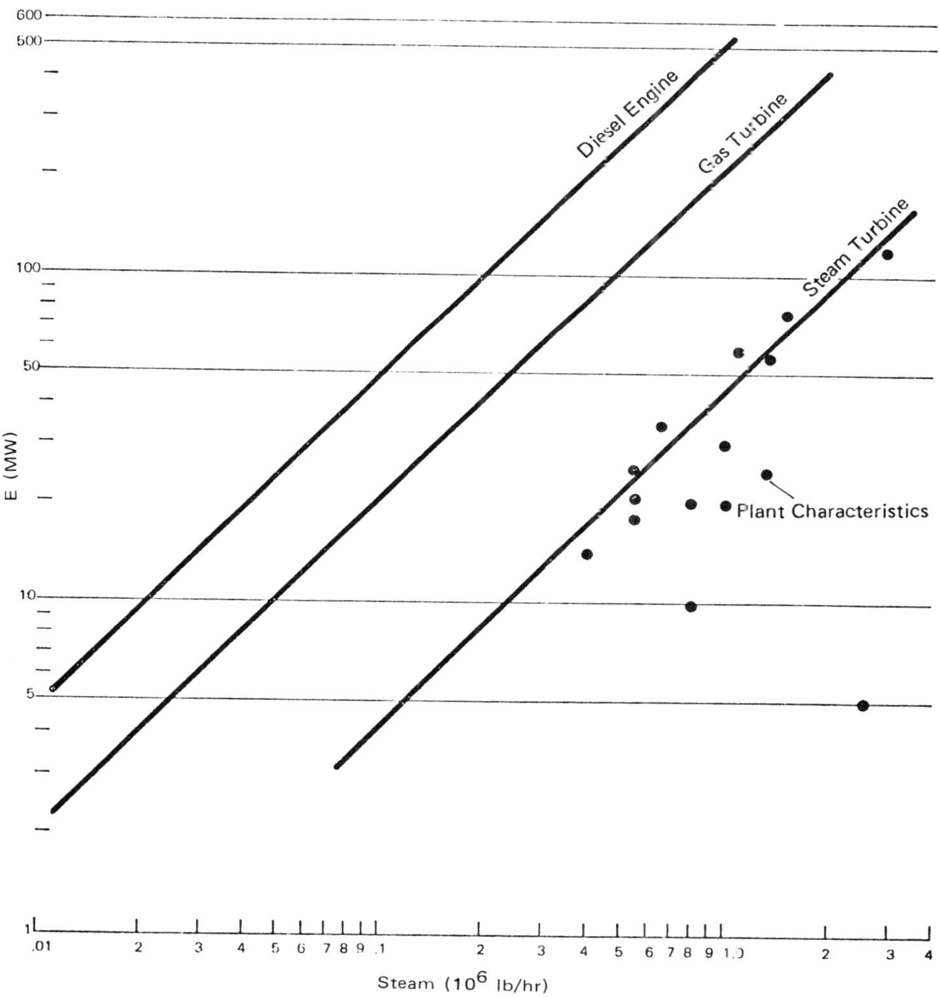

FIGURE 4.16: TOPPING OF PROCESS STEAM

(Comparison of Electrical Capacity and Steam Demand in the Paper and Pulp Industry)

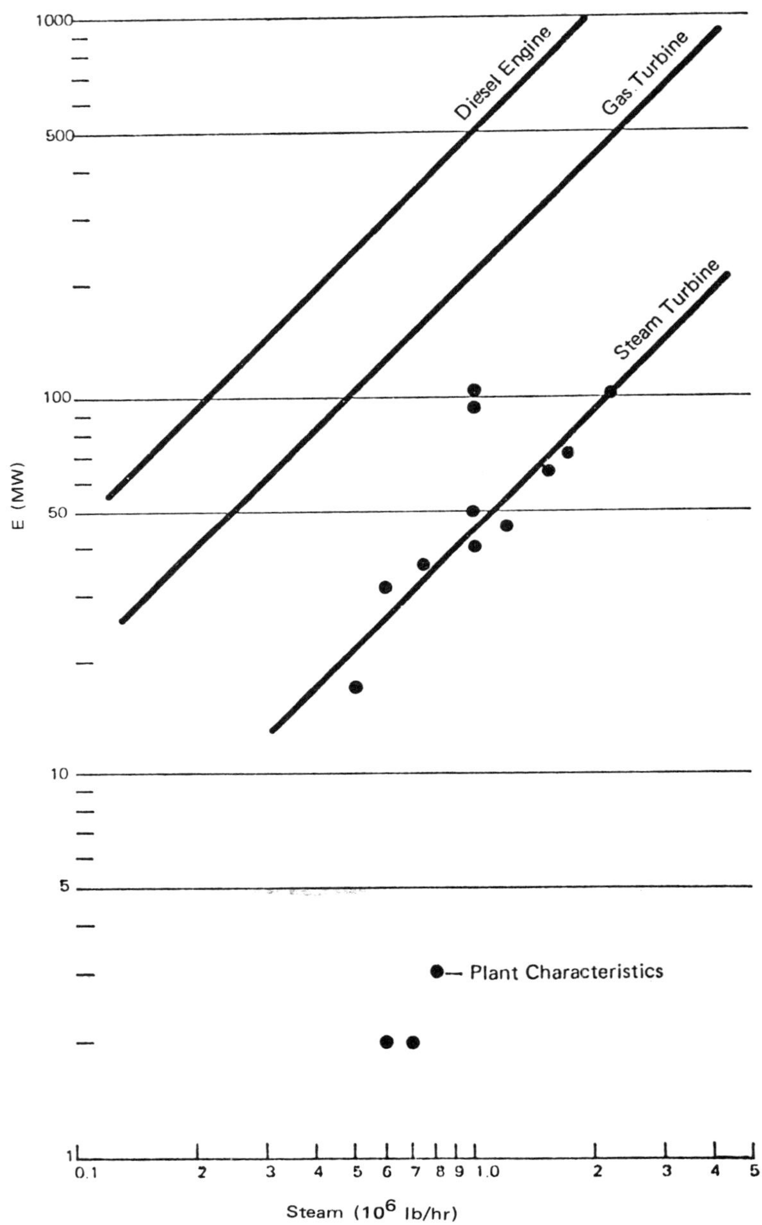

FIGURE 4.17: TOPPING OF PROCESS STEAM
(Comparison of Electrical Capacity and Steam Demand in the Textile Industry)

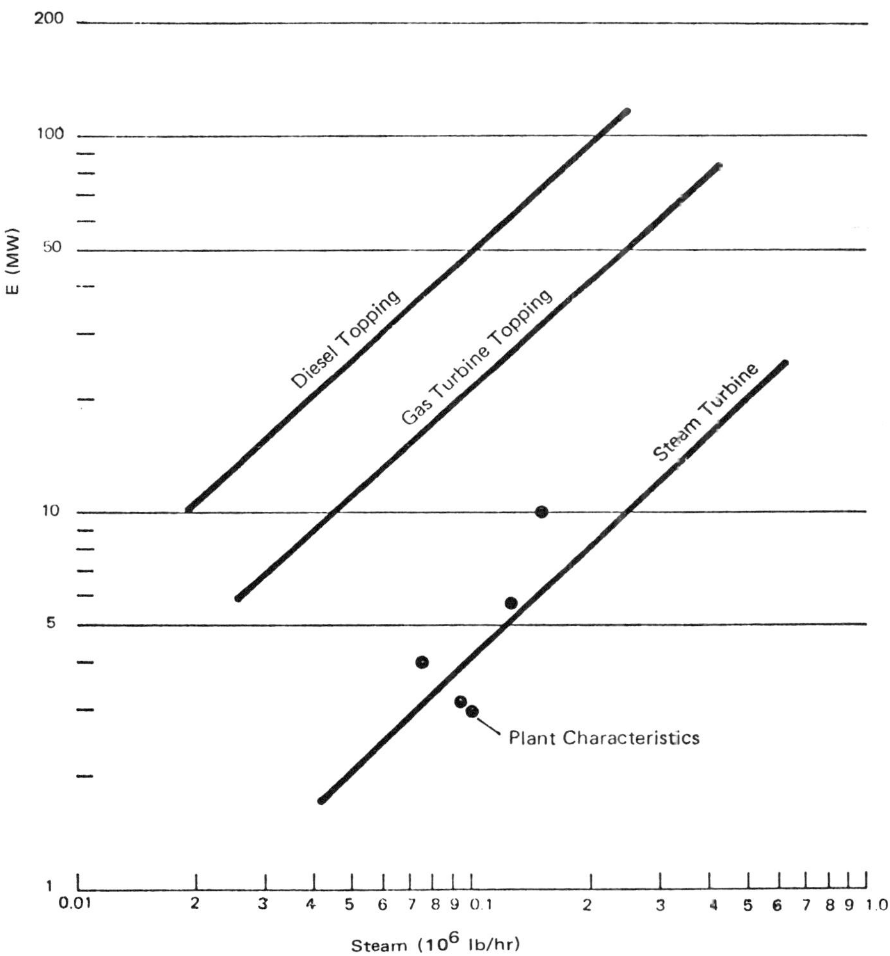

FIGURE 4.18: TOPPING OF PROCESS STEAM
(Comparison of Electrical Capacity and Steam Demand in the Food Industry)

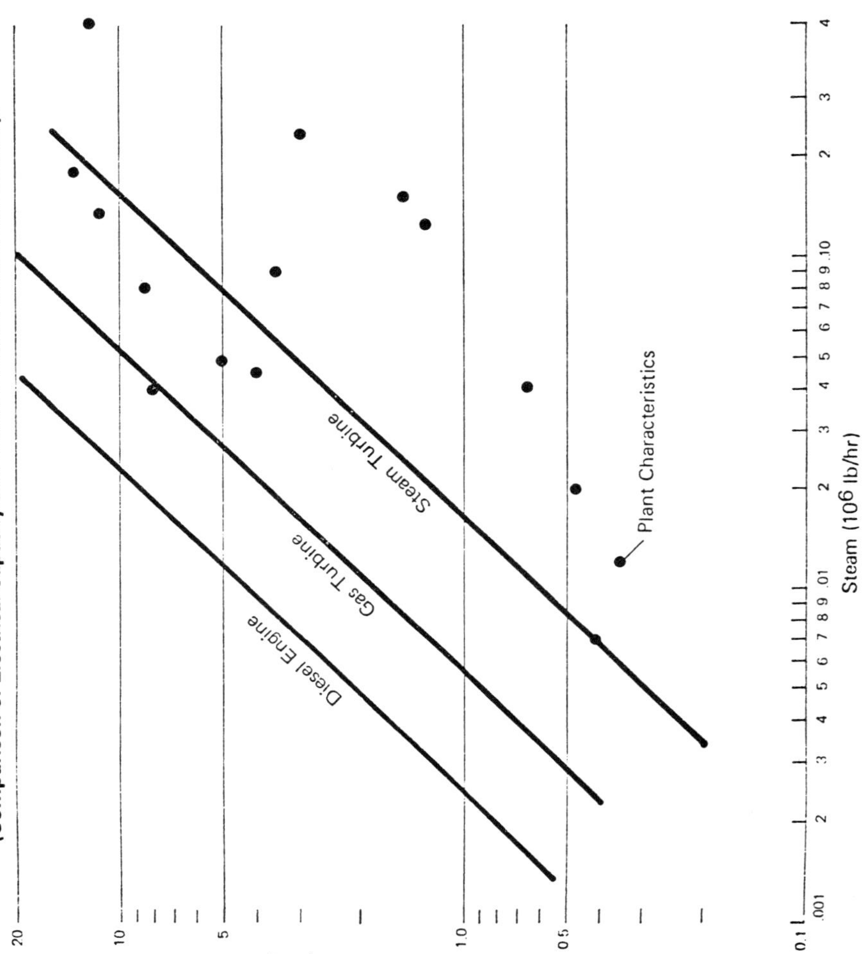

FIGURE 4.19: TOPPING OF PROCESS HEAT

(Electrical Capacity vs Thermal Energy Demand in the Chemical Industry)

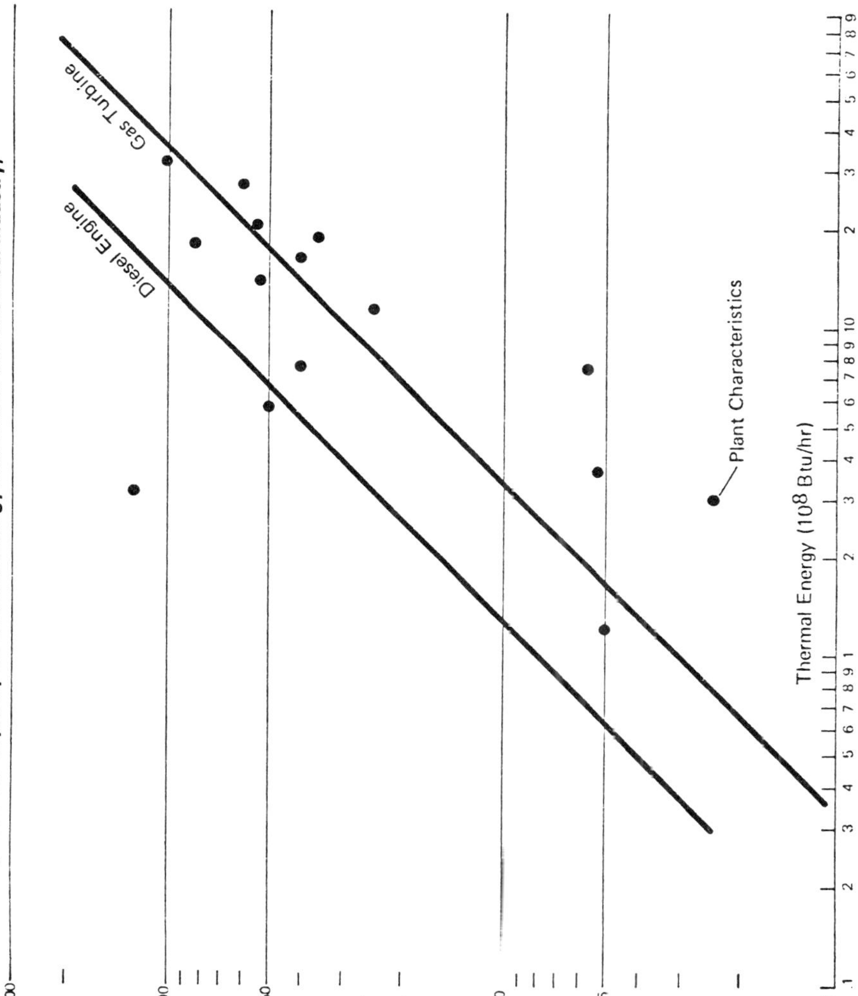

FIGURE 4.20: TOPPING OF PROCESS HEAT
(Electrical Capacity vs Thermal Energy Demand in the Petroleum Refining Industry)

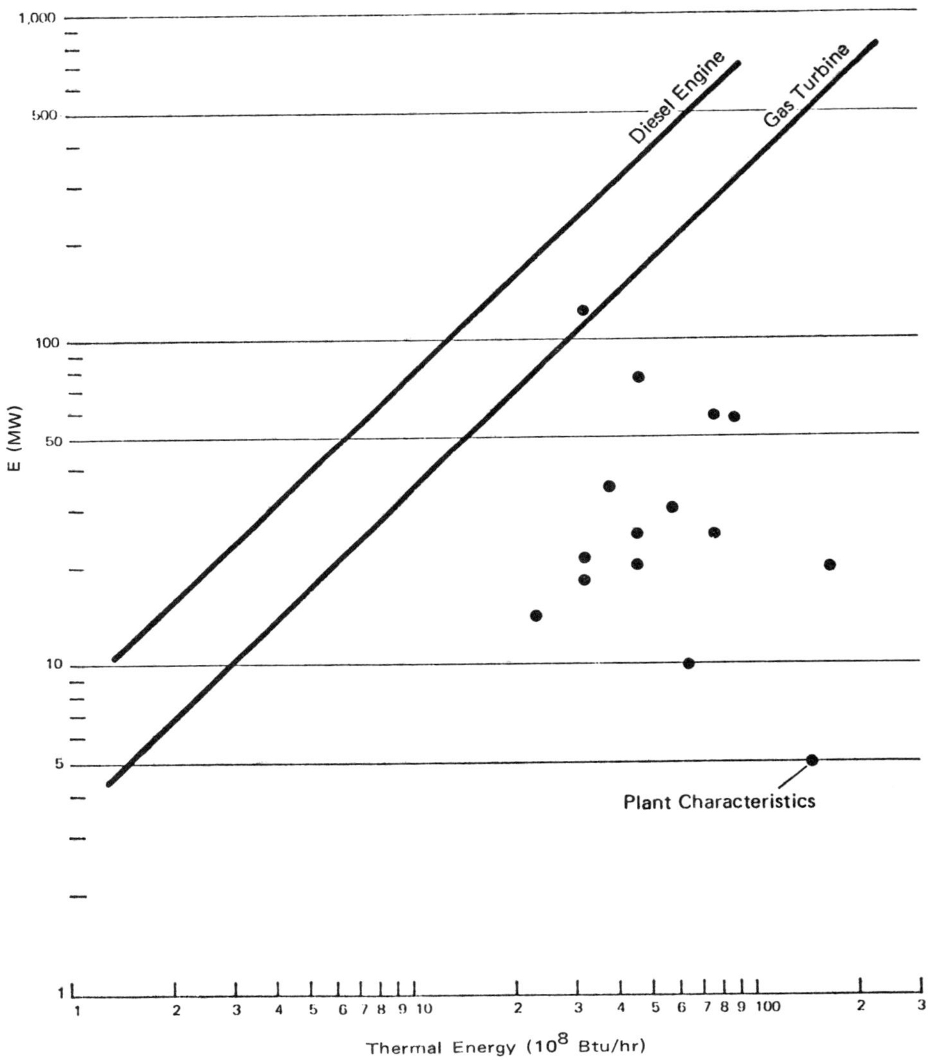

FIGURE 4.21: TOPPING OF PROCESS HEAT

(Electrical Capacity vs Thermal Energy Demand in the Paper and Pulp Industry)

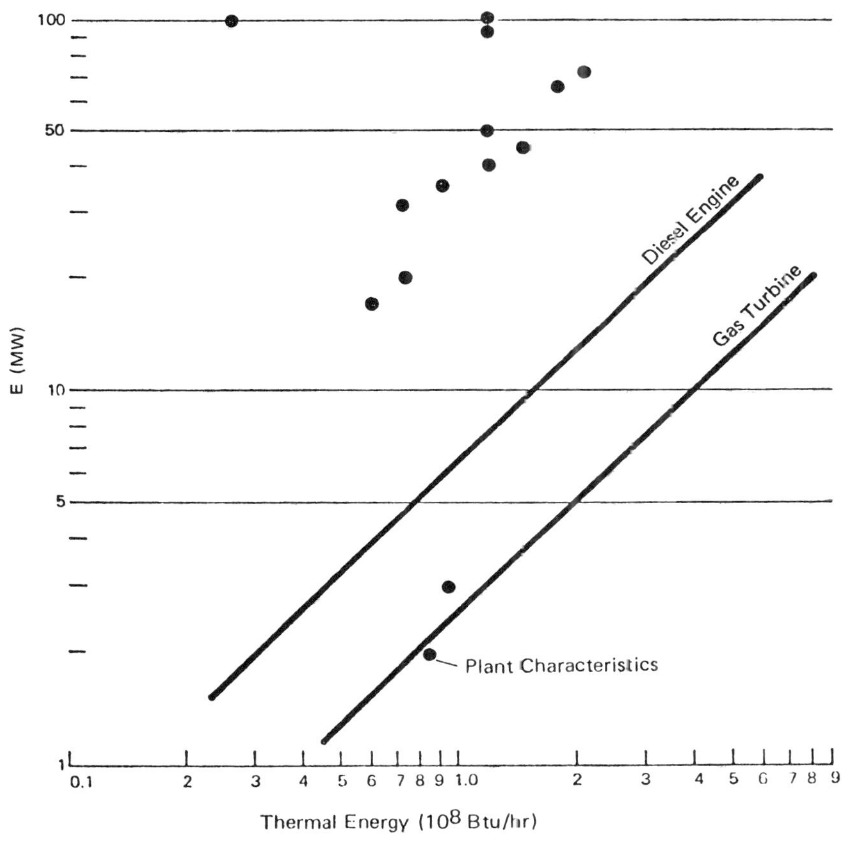

FIGURE 4.22: TOPPING OF PROCESS HEAT

(Electrical Capacity vs Thermal Energy Demand in the Textile Industry)

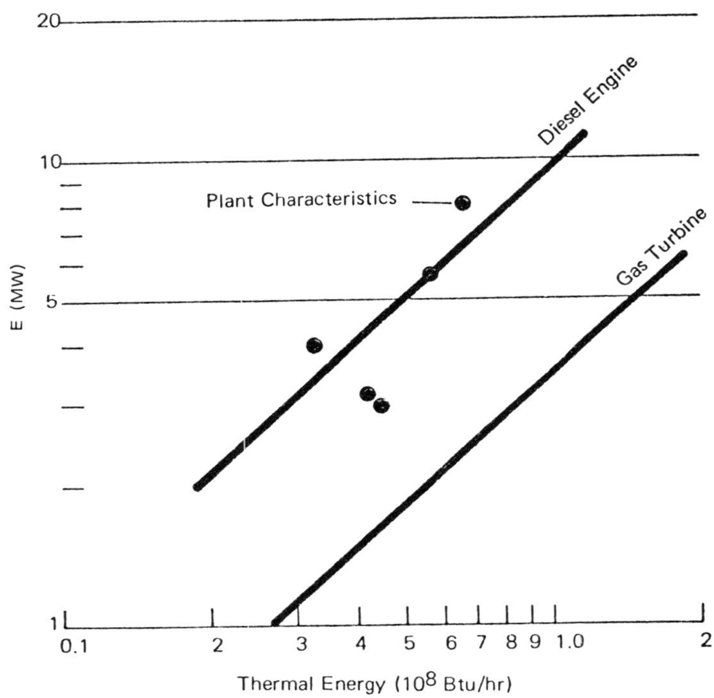

FIGURE 4.23: TOPPING OF PROCESS HEAT
(Electrical Capacity vs Thermal Energy Demand in the Food Industry)

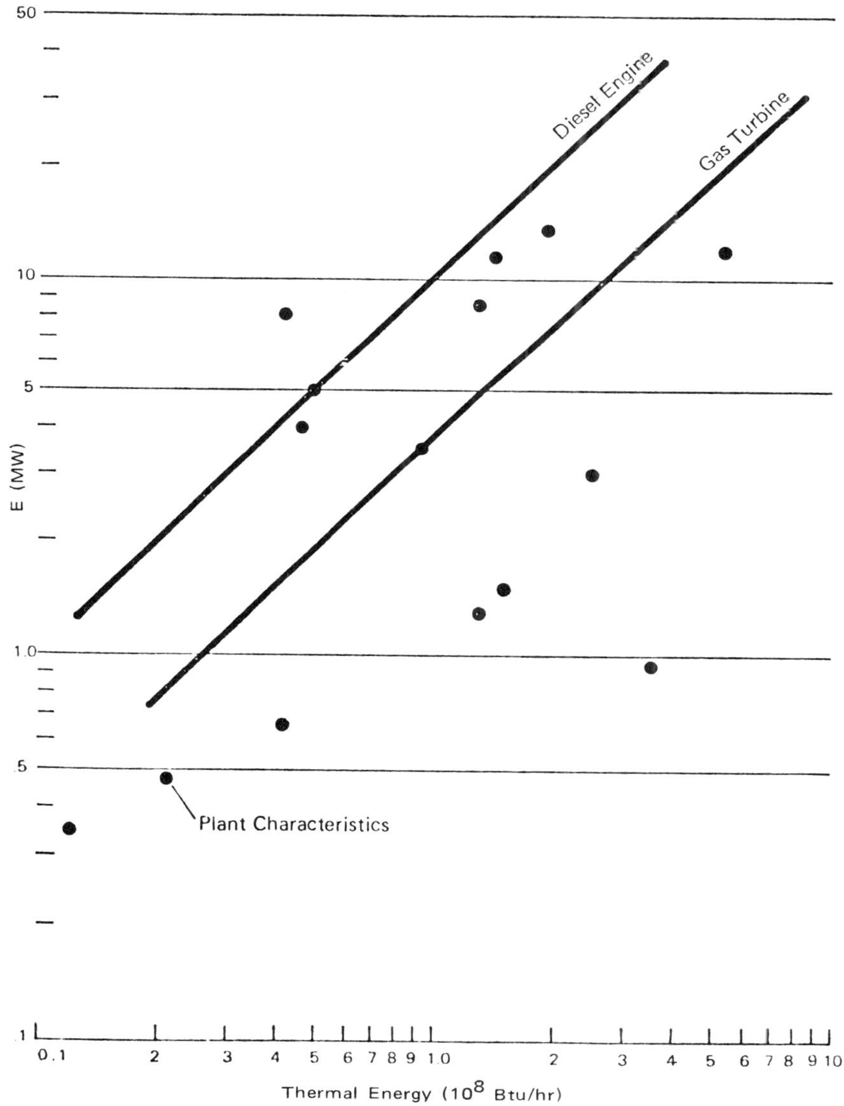

FIGURE 4.24: BOTTOMING CYCLES
(Comparison of Inverse Heat Rate and Source Gas Temperature)

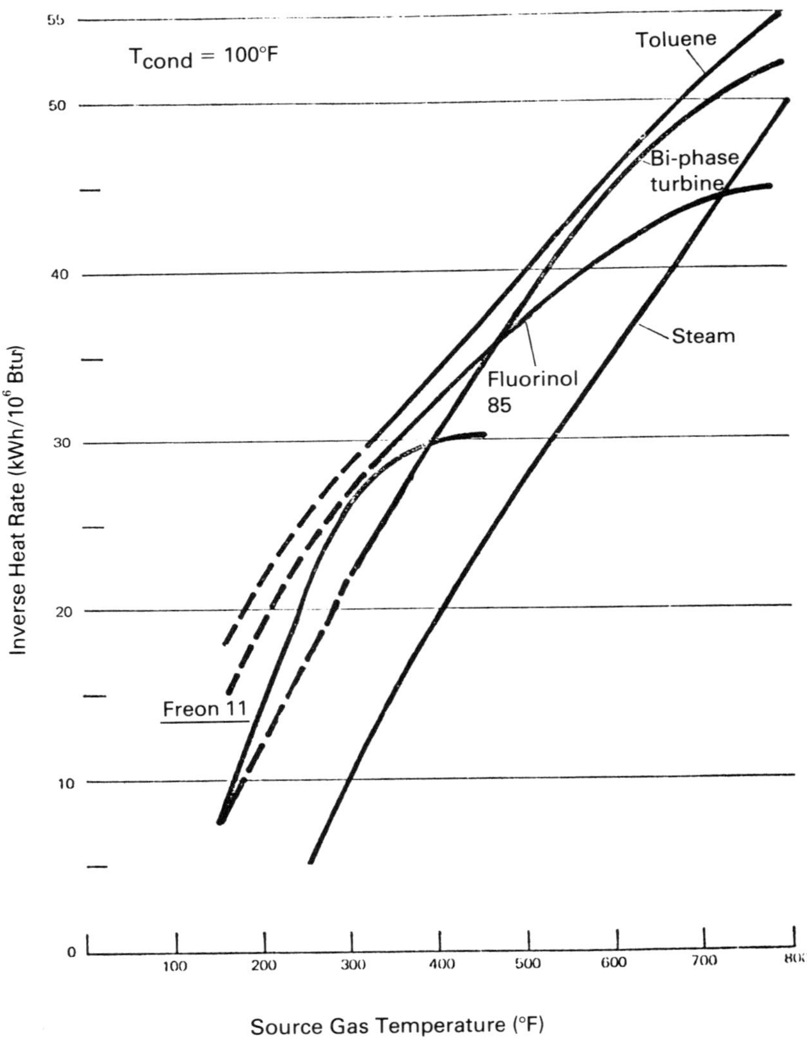

The inverse heat rates used for various temperature ranges are shown in Table 4.15.

TABLE 4.15: INVERSE HEAT RATE vs TEMPERATURE FOR VARIOUS WORKING FLUIDS

	Inverse Heat Rate kWh/10^6 Btu				
Temperature	Bi-phase turbine	Toluene ORC	Freon-11 ORC	Steam Rankine	Fluorinol-85 ORC
Less than 150°F	10	21	7	----	19
150°-300°F	17	25	20	4	23
300°-600°F	35	36	30	24	36
Greater than 600°F	52	53	30	46	45

The electricity output for each bottoming cycle was derived from an aggregation of the inverse heat rates times the waste heat profile. Outputs were plotted against total plant steam demand in Figures 4.25 and 4.26.

In both cases toluene yielded the highest output while steam Rankine was the lowest. Fluorinol 85 and the bi-phase turbine also had high performance. For the chemical industry, toluene had more than twice the output of steam bottoming. The difference is less for the petroleum industry.

These estimates are based on assumed average waste heat recovery and efficiency estimates for the industries. Specific plants may have very different waste heat profiles and any of the bottoming cycles might prove most suitable for a particular application. Nevertheless, for most applications, organic working fluids appear to be most efficient.

However, organic working fluids used in Rankine cycles and the bi-phase turbine pose serious environmental concerns that must be considered. Toluene, Fluorinol-85, Freon-11 (used in the Rankine cycle systems), and Dowtherm A (used in the bi-phase turbine) are toxic. Leakage or improper handling may be hazardous to the health and safety of workers, and improper disposal may cause environmental problems. Fluorinol-85 is a most serious hazard to workers since it is absorbed through the skin on contact. Toluene and Dowtherm A are flammable and moderately explosive. Uncontrollable release of these fluids would degrade the quality of water and possibly air. Freon-11 can break down the ozone layer of the atmosphere.

FIGURE 4.25: BOTTOMING CYCLES

(Comparison of Electrical Capacity and Process Demand in the Chemical Industry)

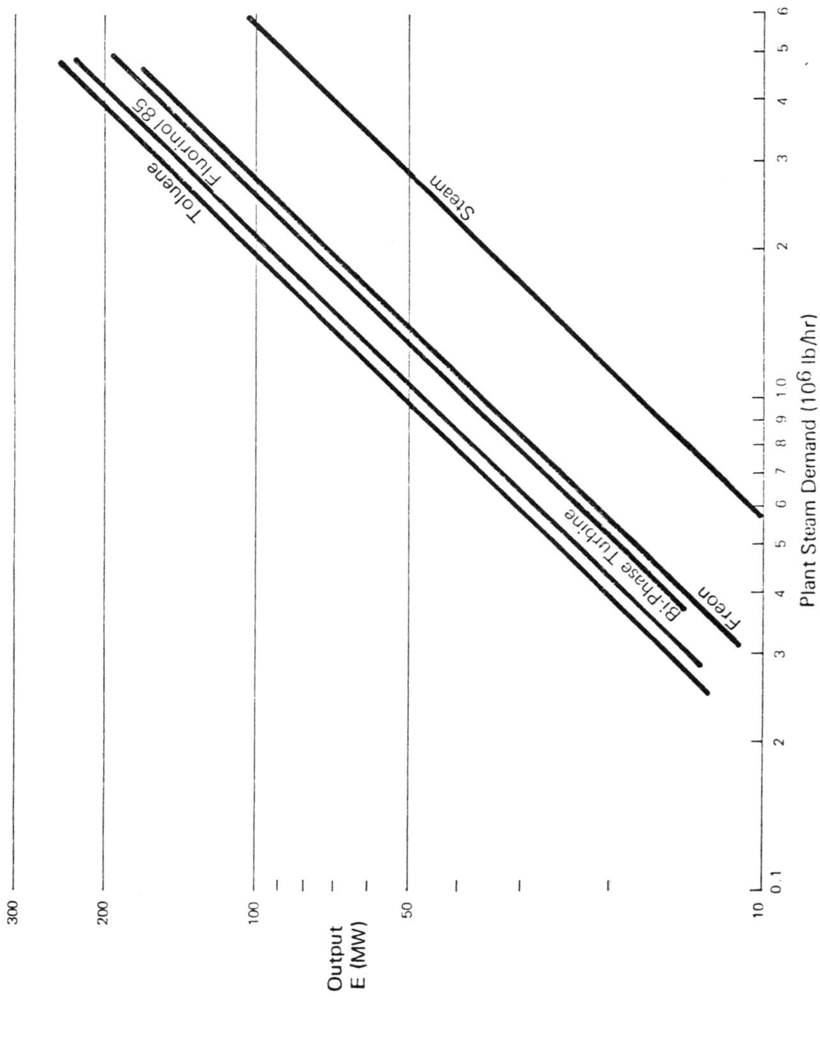

FIGURE 4.26: BOTTOMING CYCLES

(Electrical Capacity vs Process Steam Demand in the Petroleum Refining Industry)

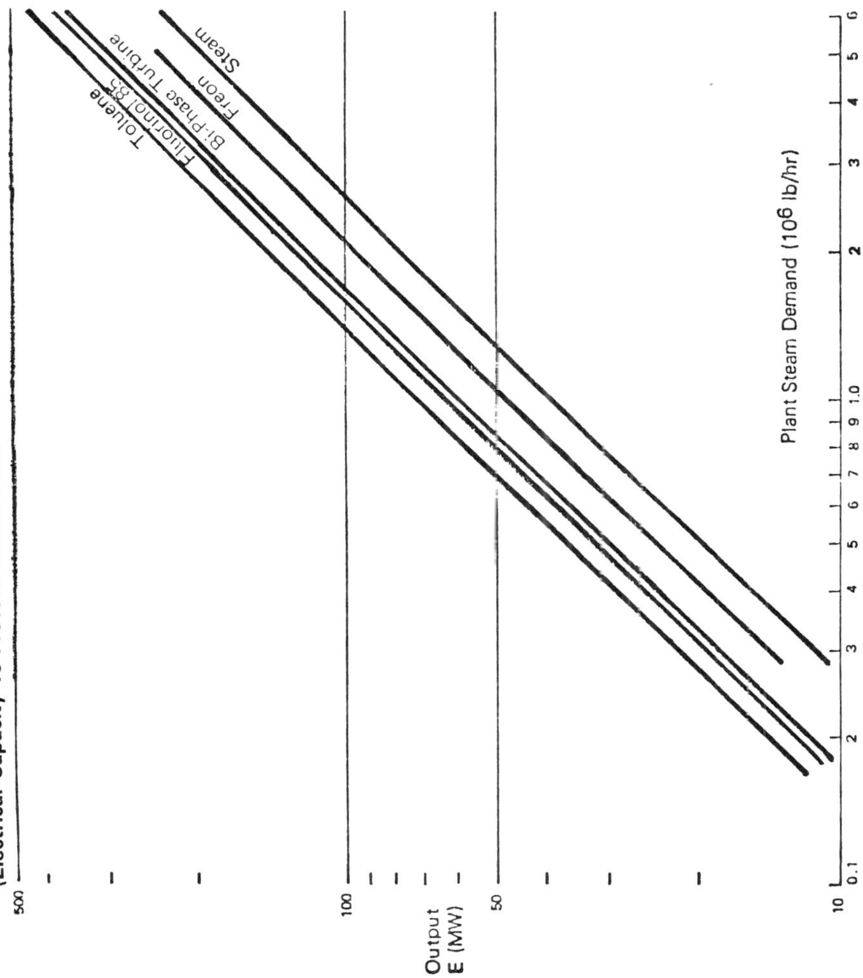

POTENTIAL SAVINGS

The purpose of this section is to estimate the potential maximum and probable energy savings that could be realized from cogeneration in five of the most energy-intensive industries. Thus, in looking for the upper limit, economic or institutional constraints were not considered.

Methodology

To establish a realistic upper limit for the topping of process steam, the estimate of the total steam used in the five industries was reduced by the amount used in ways that are unsuitable for topping. These uses, determined through discussions with industrial decisionmakers, are:

- Steam consumed in power generation (e.g., pumps, compressors)
- Process steam presently cogenerated
- Process steam produced by topping cycles
- Process steam produced from waste heat recovery boilers.

There are additional constraints that might reduce the amount of steam technically suitable for cogeneration, such as:

- Size and geographic distribution of plants
- Plant load fluctuations
- Technical expectations of management

For the purposes of this study, it was assumed that technological advances (e.g., better steam transmission and energy storage, etc.) would eliminate these constraints. In addition, it is assumed that trends towards greater use of steam (e.g., the replacement of gas dryers by steam dryers) would counterbalance any marginal change due to greater efficiencies and energy conservation efforts. Process changes and efficiencies, and conservation efforts were considered in estimating the overall growth rate to forecast 1985 values. The same growth rates in estimating 1985 values of topping of process heat and bottoming of waste heat were used.

Through discussions with manufacturers of prime movers (steam and gas turbines and diesels) net electrical output per unit of process steam (at specific conditions) was determined. By multiplying the estimated equivalent of the technically suitable process steam (in Btu per year) by the net power output ratio (kWh per Btu), the potential power generated was estimated.

To estimate the power generating potential of the topping of process heat, it was assumed that the prime mover would be either a gas turbine or a diesel engine, with exhaust supplying energy for the industrial process. It was also assumed that when the process requirement exceeded the temperature level of the exhaust stream, supplemental fuel and (in the case of diesels) air would be added. Net electrical output was determined as a function of inlet energy for each of the topping technologies. Each output ratio was multiplied by the appropriate thermal value of the temperature profile to obtain the output power of gas turbines and diesel engines. Adding these figures gave the total output power for topping of process heat.

For the bottoming of waste heat, the performance characteristics of steam Rankine cycles, organic Rankine cycles (ORCs) and bi-phase turbines were determined as a function of source temperatures. This ratio, multiplied by the appropriate thermal value of the temperature profile, yields the output power for the various bottoming technologies.

The waste heat profiles of the paper and pulp, textile, and food and kindred products industries were generally low. Furthermore, these industries already make considerable use of the waste energy. Thus, bottoming has little potential in these industries.

To calculate energy savings for all three applications, it was assumed that the cogeneration systems would be designed to satisfy thermal demand, since this was the principle most widely adopted. It was also assumed that any excess electrical energy could be exported. The following algorithms relate this potential electrical output to the potential energy savings in Btu and in barrels of oil equivalent.

The algorithm for the savings (in Btu/yr) in the topping of process steam and heat is:

$$\text{Savings} = P(10{,}000\ \text{Btu/kWh} - \text{net incremental heat rate})$$

The savings (in Btu/yr) in the bottoming of waste heat is:

$$\text{Savings} = P(10{,}000\ \text{Btu/kWh})$$

where P = kWh/yr and the net incremental heat rate (in Btu/kWh) is

4,700 for a steam turbine, 5,700 for a gas turbine or 6,500 for a diesel engine. The net incremental heat rate is an estimated average value of the five industries. It is a function of fuel type and process steam characteristics.

The algorithm to relate the potential thermal energy savings to barrels of oil per day equivalent (BPDE) is:

$$\text{Savings(BPDE)} = \text{Savings(Btu/yr)} \div 5.88 \times 10^6\text{(Btu/bbl oil)} \times 365 \times \text{LF}$$

where LF = load or capacity factor = 0.9 for all industries, except food and kindred products where it equals 0.7.

Maximum Potential Savings

Topping of Process Steam: Three types of systems for the topping of process steam were considered: back-pressure steam turbines, gas turbines with waste heat recovery boilers, and diesel engines with waste heat recovery boilers. The following characteristics for each type of system were assumed (the first set of figures given for each system being those for all industries except the food industry; the second set being for the food industry):

Steam turbines	36 kWh/10^6 Btu process steam at 850 psig and 825°F throttle condition, back pressured to 150 psig
	53 kWh/10^6 Btu process steam at 850 psig and 825°F throttle condition, back pressured to 50 psig
Gas turbines	175 kWh/10^6 Btu process steam, producing 150 psig process steam
	167 kWh/10^6 Btu process steam, producing 50 psig process steam
Diesel engine	405 kWh/10^6 Btu process steam, producing 150 psig process steam
	390 kWh/10^6 Btu process steam, producing 50 psig process steam

The amounts of steam already used for cogeneration or for other purposes that make it unsuitable for cogeneration are shown in Table 4.16. Of the estimated 3,943 trillion Btu used by the five industries in

1977, it was considered that 2,527 trillion Btu (64%) were technically suitable for cogeneration. It was estimated that in 1985, 67% of the total steam used (5,102 trillion Btu) will be technically suitable for cogeneration. Table 4.17 shows the total thermal energy technically suitable for cogeneration.

TABLE 4.16: STEAM TECHNICALLY SUITABLE FOR TOPPING OF PROCESS STEAM (10^{12} Btu/yr)

Industry	Chemical		Petroleum refining		Paper and pulp		Textiles		Food	
	77	85	77	85	77	85	77	85	77	85
Process Steam										
Total steam use	1,427	1,882	624	772	1,401	1,838	139	155	352	455
Less steam consumed in power generation	174	143	38	31	169	134	14	11	45	37
Less on-line cogenerated process steam	151	171	15	34	236	432	15	12	27	23
Less process steam produced by topping cycles	145	119	41	41	17	19	6	6	41	41
Less process steam produced from waste heat recovery	142	214	132	184		9			8	14
Steam technically suitable for topping applications	**815**	**1,235**	**398**	**482**	**979**	**1,244**	**104**	**126**	**231**	**340**

TABLE 4.17: THERMAL ENERGY TECHNICALLY SUITABLE FOR COGENERATION (10^{12} Btu/yr)

Industry	Topping of process steam	Topping of process heat	Bottoming of waste heat	Total cogeneration
1977				
Chemical	815	849	2,447	4,111
Petroleum	398	2,641	1,102	4,141
Paper/pulp	979	100		1,079
Textile	104	38		142
Food	231	205		436
Total	**2,527**	**3,833**	**3,549**	**9,909**
1985				
Chemical	1,235	1,149	3,228	5,612
Petroleum	482	2,823	1,182	4,487
Paper/pulp	1,244	102		1,346
Textile	126	37		163
Food	340	227		567
Total	**3,427**	**4,338**	**4,410**	**12,175**

The paper and pulp industry has, both proportionately and absolutely, the greatest potential for topping of process steam. The textile industry has the least absolute potential, although the proportion of its process steam technically suitable for cogeneration is high (75%). Since textiles account for only 4% of the total steam used in the five industries, the absolute potential is small.

The amount of process steam found technically suitable for topping for each industry was then multiplied by the electrical output ratio of steam turbines, gas turbines, and diesel engines to arrive at the potential electrical production. Table 4.18 shows that in 1977 in the five industries, steam turbines could produce approximately 95 million MWh, gas turbines could produce approximately 440 million MWh and diesels could produce 1,022 million MWh. In 1985, the values are estimated to be 129 million MWh, 597 million MWh, and 1,384 million MWh respectively.

This potential production of electrical energy would result in the savings shown in Table 4.19. The maximum total potential savings for steam turbines, gas turbines, and diesel engines are approximately 500 trillion Btu, 1,900 trillion Btu and 3,600 trillion Btu respectively in 1977; and 700 trillion Btu, 2,500 trillion Btu and 4,280 trillion Btu respectively in 1985. These values are equivalent to approximately 85 million to 610 million barrels of oil per year in 1977, and 116 million to 825 million barrels of oil per year in 1985 (see Table 4.20).

Topping of Process Heat: The industries studied all require extensive amounts of thermal energy for their processes. Applications range from heaters in the petroleum refining industry (high temperatures) to dryers and kilns in the food and textile industries (medium to high temperatures).

The thermal energy contained in the exhaust from a gas turbine or diesel engine could be used in the process either directly (e.g., for drying) or indirectly (e.g., by use of heat exchangers). For the purposes of this study, it was assumed that a) the thermal energy would be used indirectly, b) the heat transfer device would be 85% efficient, and c) the exhaust characteristics of the gas turbine would be those of air (because of the low fuel-to-air ratio) at 1000°F. For diesels, it was assumed that the specific heat was equivalent to that of air, and that the exhaust stream was void of combustible oxygen and at a temperature of 800°F.

On the basis of this information, net electrical output per unit of input energy was established as a function of temperature. The following

TABLE 4.18: POTENTIAL CAPACITY FOR COGENERATION
(10^6 kWh/yr)

	Topping of Process Steam			Topping of Process Heat		Bottoming of Waste Heat				
Industry	Steam Turbine	Gas Turbine	Diesel	Gas Turbine	Diesel	ORC Fluorinol	ORC Toluene	ORC R11	Steam Rankine	Bi-phase
1977										
Chemical	29,300	142,600	330,000	26,700	66,900	47,200	59,800	47,600	21,200	39,800
Petroleum refining	14,300	69,700	161,200	84,900	221,700	33,900	39,000	28,700	24,900	33,600
Paper/pulp	35,200	171,300	396,500	2,500	6,300	–	–	–	–	–
Textile	3,700	18,200	42,100	1,500	3,900	–	–	–	–	–
Food	12,200	39,400	92,000	8,100	20,700	–	–	–	–	–
Total	94,700	441,200	1,021,800	123,700	319,500	81,100	98,800	76,300	46,100	73,400
1985										
Chemical	44,500	216,000	500,200	35,200	88,500	60,700	77,100	67,700	26,100	49,800
Petroleum refining	17,400	84,400	195,200	90,800	235,000	36,300	41,600	30,700	26,700	36,000
Paper/pulp	44,900	218,000	504,600	3,000	6,500	–	–	–	–	–
Textile	4,500	22,000	51,000	1,500	3,700	–	–	–	–	–
Food	18,000	56,800	132,600	9,000	23,000	–	–	–	–	–
Total	129,300	597,200	1,383,600	139,500	356,700	97,000	118,700	92,400	52,800	85,800

TABLE 4.19: POTENTIAL THERMAL SAVINGS FROM COGENERATION (10^{12} Btu/yr)

Industry	Topping of Process Steam: Steam Turbine	Topping of Process Steam: Gas Turbine	Topping of Process Steam: Diesel	Topping of Process Heat: Gas Turbine	Topping of Process Heat: Diesel	Bottoming of Waste Heat: ORC Fluorinol	Bottoming of Waste Heat: ORC Toluene	Bottoming of Waste Heat: ORC R11	Bottoming of Waste Heat: Steam Rankine	Bottoming of Waste Heat: Bi-phase
1977										
Chemical	155	614	1,155	115	234	472	598	476	212	398
Petroleum refining	76	300	564	365	776	339	390	287	249	336
Paper/pulp	186	737	1,388	11	22	–	–	–	–	–
Textile	20	78	147	7	14	–	–	–	–	–
Food	65	170	322	35	72	–	–	–	–	–
Total	502	1,899	3,576	533	1,118	811	988	763	461	734
1985										
Chemical	236	929	1,750	151	309	607	771	617	261	498
Petroleum refining	92	363	683	377	822	363	416	307	267	360
Paper/pulp	237	937	1,766	12	22	–	–	–	–	–
Textile	24	95	179	7	13	–	–	–	–	–
Food	96	244	464	34	81	–	–	–	–	–
Total	685	2,568	4,852	581	1,247	970	1,187	924	528	858

TABLE 4.20: POTENTIAL SAVINGS FROM COGENERATION (10^6 bpye*)

Industry	Topping of Process Steam: Steam Turbine	Topping of Process Steam: Gas Turbine	Topping of Process Steam: Diesel	Topping of Process Heat: Gas Turbine	Topping of Process Heat: Diesel	Bottoming of Waste Heat: ORC Fluorinol	Bottoming of Waste Heat: ORC Toluene	Bottoming of Waste Heat: ORC R11	Bottoming of Waste Heat: Steam Rankine	Bottoming of Waste Heat: Bi-phase
1977										
Chemical	26.4	104.4	196.4	19.6	39.8	80.3	101.7	81.0	36.1	67.7
Petroleum refining	12.9	58.2	95.9	62.1	132.0	57.7	66.3	48.8	42.3	57.1
Paper/pulp	31.6	125.3	236.1	1.9	3.8	–	–	–	–	–
Textile	3.4	13.3	25	1.2	2.4	–	–	–	–	–
Food	11.1	28.9	54.8	6.0	12.2	–	–	–	–	–
Total	85.4	330.1	608.2	90.8	190.2	138.0	168.0	129.8	78.4	124.8
1985										
Chemical	40.1	158.0	297.6	25.8	52.7	103.2	131.1	105	44.4	84.7
Petroleum refining	15.7	61.7	116.2	66.4	139.7	61.7	70.7	52.3	45.4	61.2
Paper/pulp	40.3	159.4	300.3	1.9	3.8	–	–	–	–	–
Textile	4.1	16.2	30.4	1.1	2.3	–	–	–	–	–
Food	16.2	41.5	78.9	6.6	13.8	–	–	–	–	–
Total	116.4	436.8	823.4	101.8	212.3	164.9	201.8	157.3	89.8	145.9

*Barrels per year equivalent of oil.

are the weighted averages for the industries in kWh per million Btu of input energy:

	Gas Turbine	Diesel
Chemical	31	78
Petroleum refinery	32	83
Paper and pulp	27	63
Textile	40	101
Food	40	100

Table 4.21 shows the profile of input thermal energy required for the various processes. Table 4.18 shows that in 1977 approximately 124 million MWh could be generated by employing gas turbines and 320 million generated by employing diesel engines. In 1985, these values rise to approximately 140 million and 357 million MWh respectively. Approximately 90% of the total potential power generation is attributed to the chemical and petroleum refining industries.

TABLE 4.21: INPUT THERMAL ENERGY FOR TOPPING OF PROCESS HEAT (10^{12} Btu/yr)

	Chemical		Petroleum Refining		Paper and Pulp		Textiles		Food	
Industry	77	85	77	85	77	85	77	85	77	85
Process heat profile, °F										
100-600	135	184	448	480	–	–	38	37	162	179
600-1000	28	37	1,668	1,782	–	–	–	–	43	48
1000-1500	501	678	525	561	–	–	–	–	–	–
1500-2000	185	250	–	–	100	102	–	–	–	–
Total	849	1,149	2,641	2,823	100	102	38	37	205	227

This potential production of electrical energy translates to the potential savings shown in Table 4.19. The total potential savings for gas turbines and the diesel engines are approximately 530 trillion Btu and 1,100 trillion Btu respectively in 1977; and 580 trillion Btu and 1,250 trillion Btu respectively in 1985. These values are equivalent to approximately 90 million and 190 million barrels of oil per year in 1977; and 100 million and 210 million barrels of oil per year in 1985 (see Table 4.20).

Bottoming of Waste Heat: The process exhaust thermal energy (waste heat) that is presently rejected to the environment provides an added opportunity for cogeneration. Many systems such as Rankine cycles (steam and organic) can function in a topping or bottoming mode. For this study, it was assumed that these systems as well as the bi-phase turbine would be used for bottoming. Three organic Rankine cycles employing toluene, Freon, and Fluorinol were considered.

To determine the potential energy savings of bottoming, profiles of the waste heat stream of each industry were developed. As a result of discussions with industry managers and research, it was determined that of the five industries studied, only the chemical and petroleum refining industries had any real potential for bottoming. The paper and pulp, textiles, and food industries had only negligible potential. Although the overall potential for these three industries was not considered further, bottoming might be applicable at specific plants.

Table 4.22 shows the profile of thermal energy that is considered to be suitable for the bottoming systems. To estimate the potential electrical energy that could be generated from these streams of waste heat, a power conversion curve as a function of source temperature was established from various performance data. Figure 4.24 presents the results along with the assumptions used to establish a common base.

TABLE 4.22: SOURCE WASTE HEAT FOR BOTTOMING (10^{12} Btu/yr)

Industry	Chemical 77	Chemical 85	Petroleum refining 77	Petroleum refining 85
Waste heat profile				
<150°F	540	731	147	159
150-300°F	1,188	1,641	428	459
300-600°F	611	709	247	265
>600°F	108	147	280	299
Total	**2,447**	**3,228**	**1,102**	**1,182**

The bottoming potential was determined on the basis of the conversions and the profiles (see Table 4.18). It was estimated that from approximately 46 million MWh (with steam Rankine cycles) to 99 million MWh (with toluene ORCs) could be generated in 1977. The corresponding values for 1985 are approximately 53 million MWh to 119 million MWh respectively.

The potential production of electrical energy translates to the savings shown in Table 4.18. In both industrial applications, toluene organic Rankine cycles rated consistently higher than the other technologies. The steam Rankine cycle showed the least savings. The values for toluene ORCs are from approximately 1,000 trillion Btu in 1977 to 1,200 trillion Btu in 1985 and for steam Rankine cycles from approximately 460 trillion Btu in 1977 to 530 trillion Btu in 1985. The savings for the toluene ORCs are equivalent to approximately 168 million to 200 million barrels of oil per year in 1977 and 1985 respectively. For the steam Rankine cycle, savings were estimated to be the equivalent of approximately 80 million and 90 million barrels of oil per year for 1977 and 1985 respectively (see Table 4.20).

Total Maximum Cogeneration Savings Potential: In the previous sections of this chapter, the range of the maximum potential energy savings that could be realized from topping of process steam and process heat and bottoming of waste heat for the alternate technologies have been estimated. Values for the three applications were then added together to obtain the maximum potential savings for cogeneration. However, the resulting range is dependent on the mix of technology selected (i.e., varying percentages of steam and gas turbines and diesel engines for the three applications).

For example, the use of steam turbines for topping of steam, gas turbines for topping of process heat, and steam Rankine cycles for bottoming of waste heat would yield the least savings (minimum savings technology mix) while the use of diesels and toluene organic Rankine cycles would result in the greatest (upper limit). Table 4.19 indicates that the range would be approximately 1,500 trillion Btu to 5,700 trillion Btu in 1977 and 1,800 trillion Btu to 7,300 trillion Btu in 1985. Figure 4.27 shows that these energy savings are the equivalent of approximately 0.8 million barrels to 3.0 million barrels of oil per day in 1977 and approximately 1 million barrels to 3.9 million barrels of oil per day in 1985.

In terms of electrical generating capacity, these savings are the equivalent of 34,000 MW to 185,000 MW in 1977 and 41,000 MW to 238,000 MW in 1985. It is estimated that the realistic maximum will be an average of these figures. Thus, the maximum technically feasible (highly improved technology mix) would be 110,000 MW in 1977 and 140,000 MW in 1985. These values are the equivalent of approximately 1.9 million barrels of oil per day in 1977 and 2.4 million barrels of oil per day in 1985.

Probable Development

This report has focused on the technical aspects of cogeneration. Based on these assumptions, maximum potential savings for the five industries selected have been estimated. However, regulatory, institu-

FIGURE 4.27: RANGE OF MAXIMUM SAVINGS FROM COGENERATION IN EQUIVALENT BARRELS OF OIL PER DAY 1985/1977

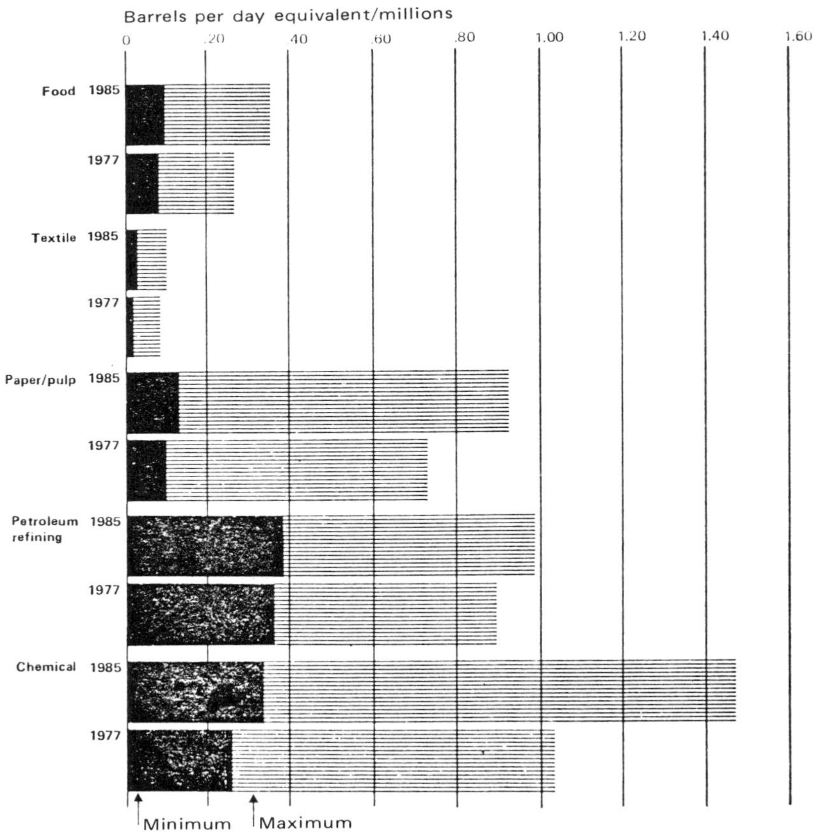

tional and other factors can significantly reduce the willingness of industrial managers to invest in cogeneration. The presence of these barriers is indicated by the fact that managers generally require higher return on investment (ROI) for cogeneration projects than for other investments.

The main regulatory and institutional barriers to industrial cogeneration arise from difficulties in interfacing with utilities, in arranging joint ventures, and in meeting environmental requirements. Government action might eliminate or reduce these barriers.

Utility Interface: There are several basic advantages to the interconnection of an industrial cogeneration system with the utility network, however, there are also problems. The basic options are:

- For the industry to purchase standby capacity to back up its cogeneration system;
- For the industry to purchase supplemental power because the cogeneration system cannot meet the entire demand of the plant;
- For the industry to sell surplus power to the utility; and
- For the industry to sell surplus power to a third party, using the utility grid for transmission ("wheeling").

However, management is concerned that standby or supplemental power contracts with utilities might not be available or might be too expensive. Furthermore, they fear that these contracts might require the cogeneration system either to deliver reserve or emergency power to the grid or to cut back power deliveries thus jeopardizing plant operations. It is not clear that industrial cogenerators will be able to obtain the price they wish from the utilities for surplus power or be allowed to wheel that power to other industrial customers. Finally, they fear that connections with utilities may bring their operations under the purview of various regulatory bodies, such as state public utility commissions, the FPC, etc.

Joint Ventures: Joint ventures between industry and utilities have numerous risks and uncertainties. Chief among them seems to be the number of parties that must be guaranteed adequate and reliable thermal and electrical energy. Since each potential participant has its own priorities and decision-making critieria, agreements would be difficult to reach.

The participation of a utility in a joint venture would facilitate construction and operation of the cogeneration system, but it might increase the likelihood of exposing the industry to unwanted regulation.

Environmental Requirements: Uncertainties about environmental policy and regulations (especially the requirements for the best available technology for pollution abatement) may inhibit the spread of cogeneration systems. This would be particularly true where requirements for siting or other environmental permits might cause delays.

Because cogeneration requires the combustion of additional fuel, local air emissions will increase. This increase should, theoretically, be offset by a net reduction in emission in the area (depending upon the types of fuel already used by the industry and by the utility that supplies it with power, and the type of fuel proposed for the cogeneration system). However, pollution control equipment may be required, particularly for systems that can burn more than one type of fuel.

Estimated Development: The following penetration ratios were used for each industry: 0.115 for the chemical industry, 0.132 for petroleum refining, 0.44 for paper and pulp industries, <0.01 for textile manufacture and 0.077 for food manufacture.

It was assumed that these ratios will not be influenced by a shift toward an improved technology mix (i.e., increased use of gas turbines or diesels). The resultant savings and electrical capacity values represent probable cogeneration potential without government incentives.

As discussed in Chapter 3, it was estimated that with a governmental program that provides for a 30% investment tax credit (ITC) clarification of FPC and Public Utility Commission (PUC) jurisdiction, an exemption from FPC/PUC jurisdiction for nonexporters of power, and an equitable purchase price for export to the grid, market development will increase by 35%.

With the probable penetration ratios, the probable electrical capacity and energy savings for various technology mixes were estimated. Figure 4.1 shows the sensitivity of electrical capacity (i.e., potential savings) to the technology mix for 1985. A 10% shift to the use of gas turbines for the topping of steam results in a 20% increase in electrical capacity over the minimum savings technology mix (from approximately 9,200 MW to 11,000 MW). With an improved technology mix, the electrical capacity would rise to approximately 20,000 MW. With a highly improved technology mix, electrical capacity should rise to 37,000 MW.

These savings are the equivalent of approximately 0.2 million barrels of oil per day for a minimum savings technology mix, approximately 0.24 million barrels of oil per day for the present technology mix, approximately 0.4 million barrels of oil per day for an improved technology mix, and approximately 0.6 million barrels of oil per day for a highly improved technology mix. A shift toward an improved technology would double energy savings for industry and the nation; a highly improved technology mix would triple these savings.

Bibliography

The following reports were used in the preparation of this book:

PB-243 823	*Energy Industrial Center Study,* by the Dow Chemical Company, Midland, Michigan, prepared for the National Science Foundation, June 1975.
PB-255 659	*A Study of Inplant Electric Power Generation in the Chemical, Petroleum Refining and Paper and Pulp Industries,* by S.E. Nydick, J.P. Davis, J. Dunlay, S. Fain, and R. Sukhuja of the Thermo Electron Corporation, Waltham, Massachusetts, prepared for the Federal Energy Administration, Office of Energy Conservation and Environment, June 1976.
RPA Reference No.: RA-77-1018	*The Potential for Cogeneration Development in Six Major Industries by 1985,* prepared by Resource Planning Associates Inc. (RPA) Cambridge, Massachusetts, for the Department of Energy, Office of Conservation and Solar Applications; Project Director, Peter Bos (RPA); Project Manager (RPA), James Williams; Project Manager, Burns & Roe, Jerome Morton; December 1977.

RPA Reference No.: RA-77-3038a — *A Technical Overview of Cogeneration: The Hardware, the Industries, the Potential Development,* prepared by Resource Planning Associates Inc., Cambridge, Massachusetts, for the Department of Energy, Division of Industrial Energy Conservation; Project Director, Howard Hagler; Project Manager, Howard Lowitt; December 1977.

MAGNETOHYDRODYNAMIC ENERGY FOR ELECTRIC POWER GENERATION 1978

Edited by R. F. Grundy

Energy Technology Review No. 20

The production of electricity by magnetohydrodynamic (MHD) processes is classed as direct energy conversion.

In an MHD generator the expanding working fluid (hot ionized gas or plasma) interacts with a magnetic field to produce electricity. The amount of current produced is set by the velocity and temperature of the ionized electrically conductive gas.

Motional electromagnetic induction (without mechanical contrivances) gives the MHD generator the unique capability among heat engines of delivering its output directly in electrical form.

The electrical power level at which MHD generators become more economical than mechanical turbines is generally considered to be over one megawatt. Success depends on the intended application and the availability of high-temperature heat sources required by MHD technology.

At this time the addition of an MHD stage to coal-fired power plants seems attractive, since the MHD stage may be able to operate directly on the high temperature products of coal combustion. Other sources of high temperatures are, of course, fusion and fission type atomic reactors.

This book is limited to the potential of MHD in the production of the much needed 60 Hz alternating current at less cost and less fuel consumption than by conventional power-generating systems. As such it presents an accurate status report based largely on federally-funded studies. A partial and condensed table of contents follows here.

ISBN 0-8155-0689-9 **230 pages**

OFFSHORE AND UNDERGROUND POWER PLANTS 1977

Edited by Robert Noyes

Energy Technology Review No. 19
Ocean Technology Review No. 6

Uncontroversial plant sites for generating electric power by any means are becoming increasingly scarce. Both nuclear and fossil fuel power plants require nearness to large quantities of cooling water and to load centers, but not on land prone to earthquakes. Consequently, utility companies are looking out to sea or beneath the surface of the earth for suitable power plant locations.

This is a most exhaustive and detailed treatise: the first six chapters being devoted to nuclear and fossil-fuel fired power plants, while the remaining six chapters deal with power generation of a more esoteric variety.

The relative costs of fixed and floating offshore plants depend on the depth of water and on the degree of earthquake resistance required. For shallow water, the fixed breakwater offshore floating station is of the most interest.

Most of the information presented in this book is based on federally-funded studies. Each chapter has its own bibliographic list, while the references at the end of the book give the full titles of the government reports on which this book is based, together with a source of their purchase. A partial and condensed list of contents follows here.

ISBN 0-8155-0680-5 **309 pages**